EFFECTIVE DATABASE DESIGN
for Geoscience Professionals

EFFECTIVE DATABASE DESIGN
for Geoscience Professionals

David R. Hoffman

Copyright © 2003 by
PennWell Corporation
1421 S. Sheridan Road
Tulsa, Oklahoma 74112-6600 USA

800.752.9764
+1.918.831.9421
sales@pennwell.com
www.pennwell-store.com
www.pennwell.com

Managing Editor: Marla Patterson
Cover and book design by Amy Spehar

Library of Congress Cataloging-in-Publication Data

Hoffman, David R.
  Effective database design for geoscience professionals / David R. Hoffman.
    p. cm.
  Includes index.
  ISBN 0-87814-828-0
    1. Petroleum--Prospecting--Databases. 2. Database design. 3. Information storage and retreival systems--Petroleum engineering I. Title.
TN271.P4 H64 2002
622'.1828'0285574--dc21

2002035541

All rights reserved. No part of this book may be reproduced, stored in a retrieval system, or transcribed in any form or by any means, electronic or mechanical, including photocopying and recording, without the prior written permission of the publisher.

Printed in the United States of America

1   2   3   4   5   06   05   04   03   02

Contents

List of Figures XI

Acknowledgments XV

Dedication XVII

1 Overview and Introduction 1
    Objectives 2
    Conventions and Nomenclature 3
    Database Misconceptions 4
        Case History: Simple Databases 4

2 Key Terms and Concepts 7
    Key Terms 7
        Database 8
        Database Management System 8
        Tables 9
        Records 10
        Fields 10
    Important Concepts 12
        Relational vs. Flat-File Databases 12
        Organization of Flat-File Databases 12
        Relational Database Concepts 13

3 The Exploration-Development Data Life Cycle 21
    Overview of Exploration and Development Cycle Phases ... 22
    Reconnaissance Phase 23
        Reconnaissance Phase Data Types and Issues 27
    Exploration Drilling Phase 31
        Acquisition and Interpretation of Exploration Drilling Data ... 32
        Shift from Surface to Well-Centric Subsurface Data ... 33
        Geophysical Data Acquisition and Interpretation 34
        Backward Integration 35
        Forward Modeling 35
    Field Delineation Phase 36
        Impact on Data Management 36
        Shift to Production and Engineering Data 37
        Data Volume Impacts 37

	Development Phase	38
	Data Interpretation	38
	Exploitation Phase	39
	Hardware Requirements	40
	Software Considerations	40
	Abandonment and Remediation Phase	41
	Data Storage Considerations	42
	Special Cases: Acquisitions and Disposals	42
	Acquisitions	42
	Disposals	43
	Data Formats and Delivery Methods	43
	Summary	44
4	**Planning Database Projects**	**47**
	Defining Project Objectives	48
	Defining Data Management Objectives	48
	Defining the Function of the Database	48
	Defining the User Interface	49
	Life Expectancy of the DBMS	49
	Defining Specific End-User Needs	50
	Case History: Scaled Application Development	50
	Involving the User in the Process	51
	Keeping the User in the Loop	53
	Continuous Improvement Processes	54
	Tailoring the Database to the Data	55
	Other Considerations	57
	Support	57
	Hardware Considerations	59
	The Database Management System (DBMS)	63
	Selection Criteria	64
	Technical Considerations	65
	Nontechnical Considerations	67
	General DBMS Types	70
	Database Applications	72
	Flexibility in Modification and Customization	73
	Ability to Link to Interpretive Applications	74
	Selecting a Data Model	75
	Case History: DBMS Selection and Customization	76
	Developing a Proprietary Data Model	77
	Selecting a Computing Platform	78
	Application-Driven Databases	79
	Application-Independent Databases	79
	Importance of Standardization	80

	Upgrade and Scalability Issues	80
	Accuracy Issues	82
	Data Transfer Issues	82
	Role of the Database Manager (DBM)	83
	Managers and Administrators	83
	General Roles and Responsibilities	84
	Integration and Coordination Functions	85
	User Communications Functions	86
5	**Data Types and Formats**	**87**
	Introduction	87
	Scalability and Portability Considerations	88
	Modifications to Commercial Data Type Definitions	88
	Data Validation and Exceptions	89
	Use Validation Rules	89
	Duplication or Redundant Data	90
	Storing Derived Data	90
	Common Geotechnical Data Types and Formats	91
	Character-Based (Text) Data	91
	Numerical Data	96
	Date and Time Data	97
	Logical Data	99
	Binary Data	99
6	**Designing the Database**	**103**
	Data Dictionaries	103
	Importance of Data Dictionaries	104
	Original vs. Derived Data	105
	What Data to Store	105
	What Data not To Store	106
	History Files and Deleted Records Files	107
	CASE Tools	108
	Definition and Application	109
	Use during Development	109
	Other Database Tools	109
	Customizing Commercial Products	110
	Data Model Extensions	112
	Generic Data Tables	113
7	**Geotechnical Data**	**119**
	Introduction	119
	Coordinate Data	120

Latitude and Longitude	120
Universal Transverse Mercator Projection Method	122
Meets and Bounds	123
Other Coordinate Systems	123
Directional Survey Data	123
Observed vs. Computed Data	124
Computed Data	128
Computational Methods	129
Depth-Related Data	131
Stratigraphic Tops, Zones, and Markers	132
Tops and Markers	132
Zones and Layers	133
Stratigraphic Exception Codes	133
Stratigraphic Nomenclature	134
Time-Related Data	135
Geophysical Data	136
Geological Age Data	138
Log and Borehole Data	138
General Organization of Log Data	139
Log Data Storage and Transfer	140
Data Editing Considerations	144
Problems of Log Data Management	145
Other Wireline Data	146
Petrophysical Data	147
Petrophysical Data from Cores	147
Indirect (Computed) Petrophysical Data	148
Derived Petrophysical Data	149
Data Management Problems of Petrophysical Data	150
Spatial Data and GIS Systems	151
Geotechnical Data and GIS Applications	151
Digital Document Storage	152
Objectives of Digital Document Storage	153
8 Data Reformatting	**159**
Goals and Objectives of Data Reformatting	159
Standardization Goals	160
Application Integration Objectives	160
Types of Reformatting Problems	161
Simple Data Manipulation	161
Complex Reformatting Problems	162
Data Conversion with Reformatting	163
Data Formatting Strategies	164

	When and When Not to Reformat	165
	Sorting vs. Indexing Data	166
	Examples of Reformatting Solutions	167
9	**Data Loading and Input**	**169**
	Same-System Data Transfer	169
	Commercial Examples	170
	Inter-Database Data Transfer	172
	ODBC and SQLNet Links to Tables	172
	Exports to DBMS Format	173
	Export to Flat File (Text)	173
	Loading from Text Files	175
	Major Import Considerations	175
	Generalized Import Procedures and Solutions	176
10	**Data Normalization**	**183**
	Definition and Importance	183
	Importance to Database Effectiveness	184
	Methods of Normalization	185
	Identifying Data Inconsistencies	185
	Search and Replace Strategies	186
	Automating the Process	187
11	**Data Validation, Editing, and Quality Control**	**189**
	Definition and Importance	189
	Methods of Validation	190
	General Validation Methods	190
	Validation of Specific Data Types	190
	Formation Tops Problems	191
	Petrophysical Data Problems	192
	Directional Survey Data	193
	Use of Geostatistical Methods	193
	Histograms and Probability Distributions	194
	Log Data Normalization Procedures	195
	Regional Statistics	195
	Database Tools	196
	Validation Tools	196
	Functions and Tools	197
	Programmatic Solutions	197
	Data Editing	197
	Editing Methods and Options	197
	Data Editing Tracking and Audit	198
	Reporting Data Problems	198
	Data Quality Control	201

Quality Control Methods 201
Quality Assurance and Documentation 201
Reviewing Data Validation Methods 203

12 Designing the User Interface 205
User Input and Feedback 205
    Planning the Interface 206
    User Critique and Feedback 208
Interface Design Options 208
    Conventional (Menu-Based) Interface Design 208
    Object-Oriented (Form-Based) GUI Design 211
    GIS (Map-Based) Interfaces 212
Validation Considerations 214
    Form-Based Validation 214
    Table Based Validation 220
    Import and Export Considerations 220
    Programmatic Solutions 221
Importance of User Involvement 221
Customizing Commercial Interfaces 222
    Support and Maintenance 222

13 Summary 225
Standardization 225
    Project Standards 225
    Documentation Standards 226
    Database System Standards 226
    Data Format Standards 226
    Interface Standards 227
User Input and Feedback 227
    User Surveys 227
    Special Advisory Groups 227
    Follow-Up Surveys 227
    Ongoing Communications 228
Documentation 228
    Project Objectives 228
    Pre-Planning Documentation 228
    Database Design 228
    User's Operational Guide 228

Appendix A Additional Resources 229

Appendix B Checklist for Geological Data Types 235

Glossary 241

Index 253

List of Figures

Fig. 1–1	Impact of data management on interpretation time.	3
Fig. 2–1	General database organization	8
Fig. 2–2	Generic database structure, showing tables within a database	9
Fig. 2–3	Individual records in a database table	10
Fig. 2–4	Data fields within a single record in a database table	11
Fig. 2–5	Schematic example of a flat–file database	12
Fig. 2–6	Relational database showing link between tables using KEY FIELD concept	13
Fig. 2–7	Relational database illustrating FOREIGN KEY concept	14
Fig. 2–8	Example of one–to–one relationship	14
Fig. 2–9	Example of one–to–many relationship	15
Fig. 2–10	Example of many–to–one relationship	15
Fig. 2–11	Using a SQL query-building interface to contruct a retrieval by "pointing and clicking"	16
Fig. 2–12	Form-based dialog	17
Fig. 3–1	The Exploration and development asset life cycle	23
Fig. 3–2	Reconnaissance phase	24
Fig. 3–3	Exploration phase	31
Fig. 3–4	Delineation drilling	36
Fig. 3–5	Exploitation phase	39
Fig. 3–6	Abandonment phase	41
Fig. 4–1	Corporate database with local and remote access	59
Fig. 4–2	Distributed or independent database configuration	61
Fig. 4–3	Multiple application, multiple database, multiple O/S solution	63
Fig. 4–4	Stand-alone database–application system	64

Fig. 4–5	Single–user data management organization	65
Fig. 4–6	Integrated, multiuser, muliple–application data management organization	66
Fig. 5–1	Well name with unpadded numeric text	93
Fig. 5–2	Sorting example using left–padded numeric text strings	93
Fig. 5–3	Example of inconsistent date format	97
Fig. 5–4	Example of consistent date formats	99
Fig. 6–1	Structure of a generic data table	114
Fig. 6–2	Generic data table with text data added	115
Fig. 6–3	Generic data table with numeric and text data	115
Fig. 6–4	SQL results for lithology data from generic table	116
Fig. 6–5	SQL results for map scale data from generic data table	117
Fig. 7–1	Location coordinates stored as latitude and longitude, angles in decimal degrees	121
Fig. 7–2	Location coordinates stored as latitude and longitude, with angles in degrees, minutes, and seconds	121
Fig. 7–3	Directional drilling measurement relationships and nomenclature	124
Fig. 7–4	Deviation survey data using only depth, inclination, and azimuth (angles in degrees)	126
Fig. 7–5	Deviation survey data using depth, inclination, and azimuth, with angles in degrees, minutes, and seconds	126
Fig. 7–6	Deviation survey data using depth, inclination, and quadrant bearing format, with angles in degrees, minutes, and seconds	127
Fig. 7–7	Example open-hole log data using the Log ASCII Standard (LAS) version 1.2	141
Fig. 7–8	Example open-hole log data using the Log ASCII Standard (LAS) version 2.0	143
Fig. 7–9	General workflow for document digitization	154
Fig. 9–1	Using the text editor to search and replace manual line breaks	177
Fig. 9–2	Initial table contents and field names	179
Fig. 9–3	New fields added to table with desired field names	179
Fig. 9–4	Update query created to move data to new fields	180
Fig. 9–5	Results of update query operation	180
Fig. 9–6	Final table contents and field names	181

Fig. 11–1	Histogram and cumulative frequency curve for data QC	194
Fig. 11–2	Prototype data problem tracking system report interface	199
Fig. 12–1	Main selection screen in an example of a menu–driven user interface	209
Fig. 12–2	Second–level menu in menu–driven user interface	210
Fig. 12–3	Computation screen in a menu–driven user interface	210
Fig. 12–4	Example of form–based validation	214
Fig. 12–5	Drop–down list box (closed position)	215
Fig. 12–6	Drop–down list box with list activated	215
Fig. 12–7	Pick–list control example	216
Fig. 12–8	Radio button control example	217
Fig. 12–9	Example of checkbox GUI control	218
Fig. 12–10	Spinbox control example	219
Fig. 12–11	Example of GUI calendar control for date entry	219

Acknowledgements

I thank the countless professionals, technicians, and support staff whose ideas, suggestions, critiques, and mentoring have guided me through a 25-year career in the petroleum industry. The importance and future impact of geotechnical data management were recognized by only a few of these pioneers in the data management field. To these individuals as well as to all those who have supported my personal efforts in this field, I express my sincerest thanks.

I also thank the editors, reviewers, and technical staff at PennWell Books for their suggestions, patience, and support during the preparation of this manuscript.

This book is dedicated to my wife, without whose support, encouragement, patience, and understanding this work would never have been completed. Every project of this nature has a unique driving force, and she has been that driving force in my life for almost three decades.

Overview and Introduction

For the nontechnical user, this book is intended to provide a solid understanding of the key elements of database design and use. Chapters are designed to be modular, so the material can be studied at a level appropriate for each reader. Even advanced users should be able to find some value in each section—if not new information, at least a fresh look at the information or an update of current knowledge and technologies. Examples are included throughout so that interested readers can use this text as a workbook in their own environments to experiment with the concepts using real-world data.

A personal computer or workstation is not required to gain knowledge from this text. It certainly helps if the reader has a fundamental understanding of petrotechnical data—how it is collected and how it is applied in the interpretation process. Fundamentally, this book focuses on the application of database concepts to solve problems in the petroleum industry.

This introductory chapter summarizes the primary goals and objectives of the book. In addition to providing a general overview of the material, it presents and discusses many important concepts and misconceptions about databases. Some general examples are shown to illustrate the critical issues and concepts.

OBJECTIVES

The audience for this book is primarily working geoscientists and managers of geotechnical professionals. Previous data management experience is not a prerequisite; rather, this text is intended to provide a general background and understanding of the concepts, definitions, and application of geotechnical data management. The best way to provide this background is through the use of specific examples and recommended solutions to common data management problems.

Conspicuously absent in this text are references to work by other authors in the field. Most geotechnical books and journal articles provide comprehensive reviews of the topic by a previous worker in the field and then offer that author's interpretation. The objectives of this text are to present my views, perspective, and experience garnered throughout my career as both a working geologist and data management professional. The lack of references is not intended to slight other workers in the field and certainly does not imply that my work is more important. Rather, my opinions, suggestions, and observations represent personal experiences that are offered to the reader as a starting point for further investigation in this important field.

Throughout the text, real-world case histories are presented to demonstrate concepts and solutions in the context of real-world examples. Theoretical solutions and concepts are of little practical value; so wherever possible a tangible example or solution has been presented.

The topic of data management in the petroleum industry has been included in countless technical presentations and journal articles over the past decade. Often, these discussions include an illustration showing how much time is wasted by interpreters looking for, loading, and editing data, and how much more time would be available if effective data management techniques were used. For example, consider the comparison in Figure 1–1:

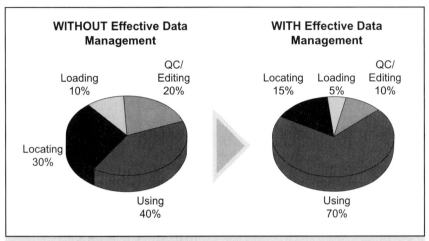

Fig. 1–1 Potential Impact of Data Management Improvements on Interpretation Time. While only qualitative, most data management professionals agree that the majority of a user's time is not spent on interpretation.

While there is little hard data available to quantify each of the above categories of time utilization (data locating, loading, quality control/editing, and utilization), most technical professionals agree that effective data management can reduce noninterpretation time by roughly half. This timesaving equates directly to more available interpretation time and is almost like doubling the interpretive staff. In today's bottom-line-oriented economy, these qualitative statistics and estimates cannot be ignored.

CONVENTIONS AND NOMENCLATURE

The subject of data management is constantly evolving, and as such the conventions, nomenclature, and definitions are a moving target. However, there are certain conventions that are used consistently throughout this book. First, there is always a great deal of variability in the use of the term *data*. The word *data* in this book always refers to multiple items. As such, *data* is considered plural. Therefore, I refer to "data are" and "these data" in all cases.

The business world is cluttered with acronyms, abbreviations, and jargon. To a certain extent, these shortcuts help simplify communication and improve clarity among those in a common field of interest. However, when faced with communicating across technical boundaries and between management and staff, these same shortcuts can create a communication impediment. I have attempted to eliminate much of the technical jargon associated with the topic of data management. Where technical terms are necessary for clarity, an entry in the glossary has been included for any obscure or confusing term. In the world of data management and information technology, acronyms are used as convenient shorthand. When a new term is introduced, the expanded acronym is always presented first for clarification.

DATABASE MISCONCEPTIONS

One of my first major assignments involving large data management projects involved an international company with massive amounts of drilling, engineering, geological, and petrophysical data. During one of the first meetings with the local staff, I was informed "all the data were already in the database." Delighted at hearing that most of the work had already been done, I eventually asked to review the database. At that point, I was escorted to a huge, musty file room containing hundreds of large three-ring binders containing hardcopy information on all of the wells, fields, and reservoirs. At first this shocking lack of technology seemed primitive and unusable. During the data collection, digitization, loading, and quality control process in the following weeks and months, it became clear that this extremely low-tech solution was a reliable, complete database that allowed me to complete the project. As discussed in the case history below, a database can be as simple as a three-ring binder or as complex as a mainframe-based database containing digital information on the worldwide assets of a multinational company.

Case history: simple databases

The international division of a midsize company in Indonesia was faced with a problem that would probably only be encountered once. A digital log database was being constructed, and legacy data tapes needed to be located, quality checked, loaded, and compared with the original paper or film versions of the logs. Unfortunately, the original data tapes were mainly nine-track standard tapes and field tapes stored in deteriorating cardboard boxes in a large storeroom in the office complex. Under the circumstances, there was no simple way of locating or retrieving a specific tape, and it was unclear exactly what data were available on tape in the first place. In addition to this being a very poor method of archival digital storage, the office space was needed for other purposes.

Many larger companies would approach a similar situation with a full-scale data management assault. This would involve commissioning a full-featured database management system, creating custom barcode, and hiring a full-time staff to index, label, and store the information on the database. In this case, however, the company recognized that the shelf life of most of the media was limited (in fact, many of the tapes were already unreadable) and that it was more important to find and load as much of the missing data as quickly as possible.

To accomplish this goal, a simple, straightforward approach was taken. First, a photocopy machine was relocated to the storeroom, and a clerk was assigned the job of photocopying the label side of each tape. As each tape label was copied, the tape was placed in a new box and the box number was noted on the photocopy. The photocopies were then placed in a series of three-ring binders, sorted by well name. As the data loading progressed, the three-ring binder database made locating the available data for any given well quick and easy (although the actual data loading was still a big problem).

This simple, low-cost solution provided a rapid answer to what could have been a massive undertaking involving several people for several months. As demonstrated by this example, a low-tech approach often provides an even more effective solution than what we perceive as a high-tech solution (which invariably is more expensive and more complex).

2

Key Terms and Concepts

Much of the confusion and many misconceptions about data management stem from an incomplete or incorrect understanding of key terms and concepts. This chapter introduces the reader to the most important data management concepts and essential terms and illustrates many of the features common to virtually all data management systems. Readers familiar with these concepts will be certain that the definitions are essentially the same as those they are familiar with, thus providing a common ground for later chapters.

KEY TERMS

Traditional data management texts tend to be jargon-laden and unduly complex for the novice. A complete summary of the technical terms, acronyms, and abbreviations common to geotechnical data management can be found in the glossary. However, some of the more important terms are defined here to provide a basis for further discussions without requiring the reader to repeatedly refer to the glossary.

Database

A database is any collection of information that provides a centralized, comprehensive reference source for users of that information. Ideally, a database is digital, is easily accessed by all users, and contains only the most accurate and complete data available. However, in the real world, databases can and do include paper documents, photographs, maps, notes, books, physical rock samples, and digital files on various incompatible computers with different formats. The goal for the data management geoscientist is to develop a database that attempts to balance practicality with the ideal database goal (i.e., all the relevant data are in one place in one consistent format).

Database management system

A data management system, as used in this text, refers to the combination of database, database management software, and related hardware and networks needed to store, manage, and retrieve data. Most geotechnical data today are available in digital format. As such, there must be some means available to access, view, and manage that information. Most database management systems provide some level of functionality to load or input data, edit, search, view, and export the information stored in the database. The level of complexity found in database management systems is directly proportional to both functionality and, in most cases, cost. This book attempts to provide general guidelines on what levels of functionality and complexity are needed for specific cases.

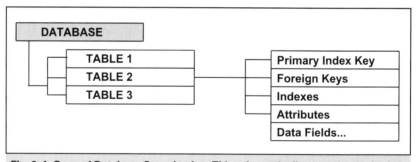

Fig. 2–1 General Database Organization. This schematic diagram shows the fundamental relationships between data tables, index keys, data, and attributes.

Tables

A database table is a collection of data records that contains common information for a particular type of data. Figure 2–1 schematically illustrates the general organization of the basic database elements. Examples of data tables might include well location information, formation tops (marker) data, and directional drilling data (see fig. 2–2). The creation of database tables is a nontrivial task, as a number of specific factors need to be considered. These considerations are discussed in more detail later in this book.

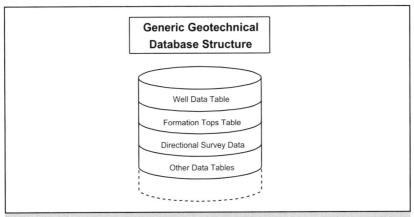

Fig. 2–2 Generic Database Structure, Showing Tables Within a Database. Most large technical database systems have hundreds of tables.

Tables in a database are organized into a collection that is referred to as a *schema*. The database schema is simply a roadmap of the tables in the database, the relationships between those tables, and the key index fields that are used to create those relationships. The schema can be presented in the form of a graphical display, or it can simply be a catalog of information about the database.

Records

The data record is the most important part of any database system because each record contains multiple elements of information that are ordered, linked, and organized with other records in an efficient manner. For simplicity, a data record is analogous to a single line in a text list or a single row in a spreadsheet (see fig. 2–3). Because data records are the most important level of organization in a database, a great deal of attention must be devoted to ensuring they contain all necessary information while excluding unneeded or duplicate information. These design criteria are collectively referred to as *normalization*.

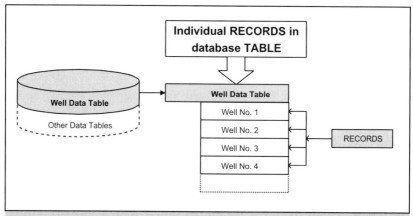

Fig. 2–3 Individual Records in a Database Table. In this illustration, the well data table contains multiple records, each of which contains data on a single well.

Fields

Data fields are the individual data elements that, together, make up database records (Fig. 2–4). Each field contains a single data item and has a specific data type. Database fields can be numeric, text, logical, dates, or binary, with specialized subtypes for each. Each of the main data types is discussed in more detail in the *Data Types and Formats* chapter.

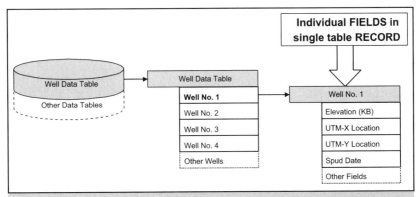

Fig. 2–4 Data Fields Within a Single Record in a Database Table. Following the previous example, a single record in the well data table contains multiple data fields that contain the detailed information about that well.

The individual data fields in a database record can be grouped into two general categories:

Data fields are the basic building blocks of a data record. They are generally related to one another in a single record and are organized in various ways. Data fields can include virtually any type or format of data, including numeric, text, date, logical, and binary formats.

Key fields or index fields are fields used to relate records in one table to records in other tables. Key fields can be *primary* keys or *foreign* keys. A primary key is a field that uniquely identifies a record in a single table. There can be no duplication of a primary key in a table. Records can also contain foreign keys, which are used to link information to data stored in other tables (see detailed discussion of foreign keys later in this chapter). Records in a table can contain duplicate foreign key values, but each foreign key must be a primary key in another table. Key fields can contain real data values or can simply be an index that only provides a link between tables.

Important Concepts

Relational vs. flat-file databases

Most geoscientists view any collection of information as a database. A database can be a sophisticated electronic data repository or something as simple as a three-ring binder filled with handwritten notes. In this book, we examine only computer-based databases, which include a wide variety of types. The two types of databases used widely today are the relational database and the flat-file database. Relational databases are commonly more sophisticated and more extensive, while flat-file databases are more limited and generally are used by a fewer individual users or small workgroups.

Organization of flat-file databases

A flat-file database is a generic description of any computer data file that exists in isolation and is not dependent on other files or a particular data management system to be useful. Figure 2–5 shows a hypothetical flat-file database of well and formation top data. Although there are a number of fields that are duplicated in multiple records, the data contained in the flat file are "self-contained" and do not require information from other tables to be useful. Some flat-file databases consist of an organized set of text files that individually may or may not conform to a specified style or format. Examples of this type of database could include directional drilling information (one file per well), seismic shotpoint location or navigation data, formation top data, or virtually any other type of geotechnical data that can be organized by well, line, or other logical subdivision.

Well Data Table					
Well No. 1	KB Elevation	Location UTMX	Location UTMY	Top 1 Depth	Other Fields
Well No. 1	KB Elevation	Location UTMX	Location UTMY	Top 2 Depth	
Well No. 1	KB Elevation	Location UTMX	Location UTMY	Top 3 Depth	
Well No. 2	KB Elevation	Location UTMX	Location UTMY	Top 1 Depth	
Well No. 2	KB Elevation	Location UTMX	Location UTMY	Top 2 Depth	
Other Records					

Fig. 2–5 Schematic Example of a Flat-File Database. In this example, each record contains all the information needed to locate the well and a specific formation top. However, note that there is a great deal of repetition in the data fields for each well.

Flat-file databases can also include spreadsheet files, although most modern spreadsheet programs allow the creation of links between multiple worksheets, as well as other fundamental features common to relational database systems.

Relational database concepts

A relational database consists of multiple data tables in which the individual records are related, or linked, to associated data in other tables by a series of unique key data elements. These data tables are all contained within a single data management environment and are managed by the data management software application. The internal structure of the tables and data elements is defined and maintained consistently throughout the total system.

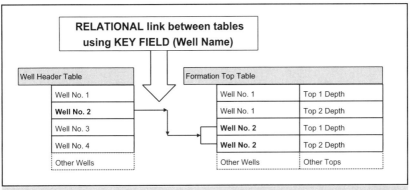

Fig. 2–6 Relational Database Showing Link Between Tables Using KEY FIELD Concept. In this example, the key field (well name) is used to establish a link between the basic well data in the well header table and the formation tops data in the formation top table.

Because the information in a relational database is distributed among multiple tables, it is necessary to make links or relationships between the tables using a series of key fields. A *primary* key field is unique within a table, i.e., no duplicates are allowed in primary key fields. However, a table can contain many *foreign* key fields. *Foreign* key fields are unique in another table and provide a means of linking data from other tables. Foreign keys do not have to be unique within a table.

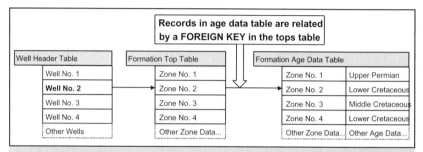

Fig. 2–7 Relational Database Illustrating FOREIGN KEY Concept. This example shows how the foreign key field (in this case, zone name) is used to link the formation top table to the formation age data table.

The fundamental relationships that can exist between tables in a relational database include the following types of links or relationships. In this discussion, parent table refers to the table from which the link is made. The child table is the related table.

One-to-one relationship. The one-to-one relationship is the simplest link that can be made between fields in two tables. The information in a field in one table is linked to one (and only one) field in another table. These links require that the key fields being linked are unique—that is, there can be no duplicates in either table. An example of a one-to-one relationship is illustrated in Figure 2–9. In this example the well information master table is linked by a one-to-one relationship to a table of well elevations.

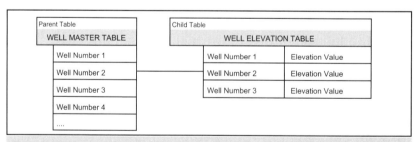

Fig. 2–8 Example of One-To-One Relationship. In a one-to-one relationship, one record in a table is related to only one record in another table using a key field.

One-to-many relationship. The one-to-many relationship relates the unique value in the parent table to more than one record in the child table. An example of this type of relationship could be a well location table, where one well is linked to a stratigraphic tops table where (presumably) multiple entries or records exist that all relate to the single well in the parent table.

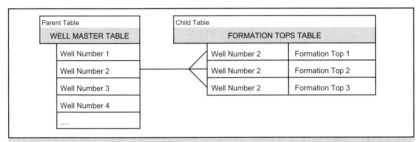

Fig. 2–9 Example of One-to-Many Relationship. This type of relational link connects the information on a single record in one table (in this case, the basic well header data) to multiple records in another table (in this example, formation tops).

Many-to-one relationship. The many-to-one relationship is essentially the reverse of the one-to-many relationship in that multiple records in the parent table are linked to a single record in the child table. An example of this type of relationship could be a well location master table that is related to a platform information table using a many-to-one relationship (i.e., multiple wells are drilled from the same platform). Using this type of relationship would allow all wells on the platform to be related to the single unique entry in the platform information table.

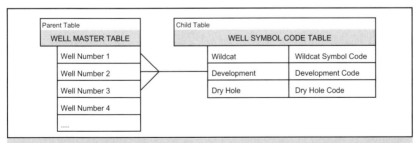

Fig. 2–10 Example of Many-to-One Relationship. The one-to-many relationship provides a link between multiple records in one table (here, the well header data), with a single record in another table (in this case, the well code for a specific type of well).

Retrieving data from tables. The retrieval and use of data from a database is, of course the ultimate goal of the data management process. For relational databases, the retrieval process is handled using Structured Query Language (SQL, pronounced See-Qwel) commands. Although the format of SQL is suspiciously like a programming language, it is actually a procedural command language. The basic use of SQL only requires a very limited subset of commands and is easy to learn. Furthermore, many applications now have built-in SQL interfaces that allow the user to simply point-and-click on the desired data without actually writing any SQL statements. Often, this takes the form of a query-by-example (QBE) interface. A QBE interface is a simple method of building database queries without knowing any of the fundamental SQL commands that actually do the work. It is important to remember that all data are retrieved from a database using SQL statements at some point in the process.

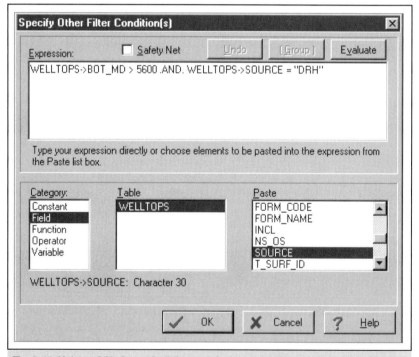

Fig. 2–11 Using a SQL Query-Building Interface to Construct a Retrieval by Pointing and Clicking. This example illustrates a typical SQL language interface, where the user "drags and drops" commands and keywords to construct a valid SQL query.

Key Terms and Concepts

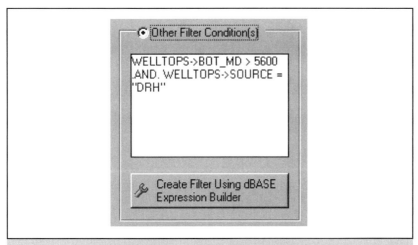

Fig. 2–12 Form-based Dialog, Allowing Direct Entry of SQL Commands or Access to the Expression-building Interface. This interface displays the SQL command syntax and allows direct input of SQL commands (by advanced users), but will still access the interactive command interface shown in the previous example.

Views (multiple tables). Views are predefined queries that incorporate all necessary tables, relational links, and SQL retrieval statements. What the user sees is a selected subset of the fields in the various tables that provides the user with a specific view of information stored in the database. While views can be created by the user, knowledge of the query language and various database utilities makes this more the realm of the advanced user or developer. Views can be extremely useful, in that they provide a simple method of getting routine data from the database in a predefined format without resorting to developing SQL statements every time the data are to be retrieved.

Data export functions. Most relational databases have some sort of data export functions. These, combined with query operations and reporting functions, are the primary methods of extracting data from the database for use in other applications and hardcopy reports. A data export function can be a dump of all the contents of all the fields in a selected table or view or a customized subset of data in a specific format that can be easily imported into another application. In many cases, these custom export formats require programming support, but a rudimentary knowledge of SQL retrieval methods allows very sophisticated data retrievals and exports.

Reporting functions. Despite efforts to move toward a paperless business environment, hardcopy reports, summaries, and listings are still necessary in many cases. Database applications normally have some sort of reporting application, either incorporated into the application or available as a compatible optional add-on. In either case, report-generating software is indispensable for creating custom data listings, routine data summaries, and other essential hardcopy documentation.

Comparison of database types

The most fundamental difference between relational and flat-file databases is that the relational database connects associated data elements in multiple tables using a series of key fields. In comparison, the data files contained in flat-file databases are each a complete and stand-alone collection of data elements that do not rely on other tables or the data management system to be useful. In many cases, data files in a flat-file database file are self-defining. That is, the file contains all information (usually in a designated file header area) that specifies the data elements contained in the file and format in which they are stored. An excellent example of a popular and industry-accepted self-defining file is the Log ASCII Standard (LAS) file used in the logging industry (see *Geotechnical Data* chapter). A relational database, on the other hand, usually requires some knowledge of the database schema, a query engine, and query language to extract useful collections of data.

To best illustrate the differences between flat-file and relational databases, Table 2–1 compares the features and functionality of each type.

Table 2–1 Comparison of Flat-File and Relational Databases

	Flat-File Database	Relational Database
Data Access Time	Access requires an application script that opens the file, searches the entire file for a particular item, and returns the result of the search.	Using SQL, individual data items can be found very quickly and efficiently
Loading Large Data Volumes	Individual files can be bulk loaded quickly and easily with the appropriate loading scripts.	In most cases a data table view is used to organize the data; then an export file can be generated which contains the desired data.
Data Currency	File contents are only as current as when the file was created. Any changes to the original data after that time are not reflected in the file.	Data tables always contain the most current data, as editing and new data entry are done directly on the contents of the relational tables.
Data Editing	Editing must be done through an application that loads the current file contents and then rebuilds the file after edits, additions, and deletions are completed.	New data, deletions, and modifications are done interactively with the relational tables.

Now that the basic concepts of database management have been discussed and key terms and nomenclature have been defined, the next chapter will show the changing role of data management in the exploration-development data life cycle. By examining the data-related activities for a petroleum asset from reconnaissance to exploration, development, and abandonment, the relative importance of data management can be illustrated with tangible examples.

3
The Exploration-Development Data Life Cycle

Geotechnical data are the most critical ingredient in any exploration or development project. As such, data management can and must play a pivotal role in the life cycle of any exploration or development project. Data management affects all phases of exploration and development. The need for accurate, complete, and well-organized data is critical to the success of any project of this type. With the ever-increasing competitive environment of today's oil industry, lower-margin operations require leaner staff levels, lower budgets, and fast asset cycle time. Effective data management is the key to maintaining all of these factors in balance.

The objective of this chapter is to present the reader with a general overview of the exploration-development cycle and the interaction of data management issues during each phase of this cycle—exploration, development, and disposal. From the earliest identification of an area's exploration potential to the abandonment of the last producing well, data management plays a critical role in all facets of the operation. While there are common problems and solutions throughout the cycle, each phase creates unique data management problems.

This chapter illustrates key concepts in geotechnical data management by following the evolution of a typical oil and gas asset through its entire life cycle. In each of the major phases of this cycle, the key data management concepts appropriate for that phase will be illustrated. This exploration-development life cycle requires a constantly changing evolution of data management needs and methods. In each phase, there are different levels of data quality, data quantity, and data accuracy.

OVERVIEW OF EXPLORATION-DEVELOPMENT CYCLE PHASES

An oil and gas asset passes through six major phases during its life cycle. While these terms are broadly defined and have very gradational boundaries, the major phases of the exploration-development cycle remain the same:

- Reconnaissance
- Exploration drilling
- Field delineation
- Development
- Exploitation
- Abandonment and remediation

The total cradle-to-grave period of this cycle generally lasts much longer than the typical career span of most geoscientists—or, for that matter, most companies (particularly with the massive restructuring and mergers of the past decade). As a result, most properties or assets have multiple generations of both owners and interpreters. Thus, two special cases of the exploration-development cycle—acquisitions and asset disposals—are discussed separately at the end of this chapter.

The exploration-development asset life cycle matrix (Fig. 3–1) illustrates the general importance of various data types during the phases of the life cycle. Each phase of the cycle is discussed in the following sections, with illustrations of the key data management concepts and methods.

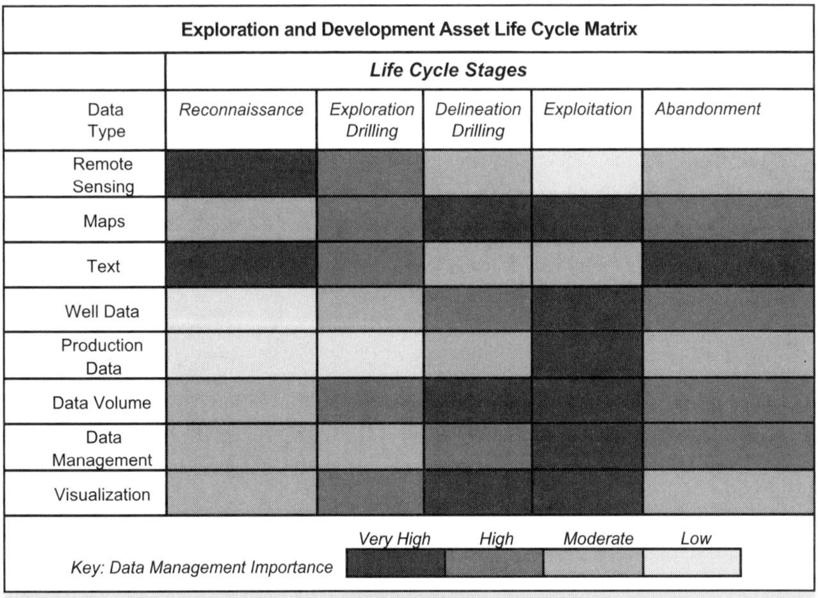

Exploration and Development Asset Life Cycle Matrix					
	Life Cycle Stages				
Data Type	Reconnaissance	Exploration Drilling	Delineation Drilling	Exploitation	Abandonment
Remote Sensing					
Maps					
Text					
Well Data					
Production Data					
Data Volume					
Data Management					
Visualization					

Key: Data Management Importance — Very High, High, Moderate, Low

Fig. 3–1 The Exploration—Development Asset Life Cycle. The qualitative, relative importance of major categories of geotechnical data types is shown for each phase.

RECONNAISSANCE PHASE

The earliest part of the exploration-development cycle begins with recognizing (or speculating) that an unexplored or underexplored geographic area is a commercially viable prospect for oil and/or gas exploration. Most often, this recognition is based on older surveys and studies, maps, surface features, imagery, and limited well control data. At this point, there is very little hard, or quantitative, data to work with. Most of the available data are textual, and they most often are surface dominated (as opposed to subsurface dominated). Figure 3–2 shows the major geotechnical data types and their relative importance during the reconnaissance phase.

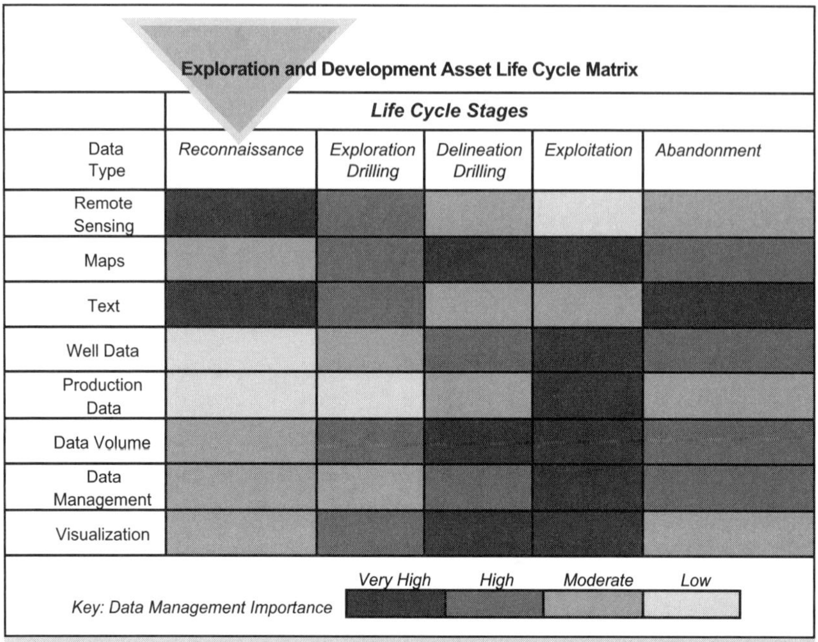

Exploration and Development Asset Life Cycle Matrix					
	Life Cycle Stages				
Data Type	Reconnaissance	Exploration Drilling	Delineation Drilling	Exploitation	Abandonment
Remote Sensing					
Maps					
Text					
Well Data					
Production Data					
Data Volume					
Data Management					
Visualization					

Key: Data Management Importance — Very High, High, Moderate, Low

Fig. 3–2 Reconnaissance Phase of the Exploration – Development Life Cycle. The relative importance of major geotechnical data types is highlighted.

Textual data, in the form of written reports, maps, notes, and other hardcopy data, are some of the most difficult types of geotechnical information to manage. Of course, during this phase they are often the only data available, which means that special attention must be devoted to data management. In most cases, these data must be shared by various disciplines, possibly located in disparate geographic locations. Early in the project life cycle, someone must determine whether the paper documents will be scanned and stored in a document-storage application, scanned and converted to text in a database, or captured using some combination of these two methods (see *Digital Document Storage* section in *Geotechnical Data* chapter).

Reconnaissance data are predominantly map-focused surface data rather than well-focused subsurface data, which dominate the latter phases of the exploration-development life cycle. Maps, cross-sections, well schematics, and other paper-based graphical data are highly variable with regard to scale, geographic coverage, content, format, and even language. In most cases, ownership or source of the data may be difficult or impossible to determine. Rarely, if ever, will hardcopy data used for reconnaissance work be available in digital format because of the vintage, type of data, or data source (trade, previous acquisition, etc.). Like written data, one must determine early in the project life cycle whether the map/graphical data will be digitized, scanned, retained as paper prints, or saved in some combination of these methods.

In the case of either type of data, comprehensive indexing of the actual information contained in the hardcopy report, map, or other paper data is the most critical and fundamental part of being able to locate information from multiple sources, in multiple formats, or from different vintages. This content index information is generally referred to as *metadata*. While there have been many recent advances in metadata collection (in some cases provided by the data source or vendor), metadata collection is still a time-consuming, tedious, and meticulous task that must be conducted by knowledgeable, experienced professionals.

Some of the key questions that need to be addressed when developing a metadata indexing system are:

What geographic area does the data encompass? Virtually all geotechnical data are earth-centric, and spatial data in some form are essential to understanding and using the data. Geographic metadata are often captured in two forms: quantitative and qualitative.

Quantitative geographic metadata, in the form of latitude-longitude or geographic coordinates, are needed if the data need to be "searchable" via some sort of query language (see *Coordinate Data* section in *Geotechnical Data* chapter). Latitude-longitude locations are absolute references and are the best choice for referencing a geographic area. Geographic locations (relative x-y locations, such as the common UTM system) require that additional information be captured, including the projection method and all constants needed to convert the x-y coordinates to absolute (latitude-longitude) coordinates.

Qualitative location data are generally textual references to the basin, area, offshore block, etc. Because of the variable nature of qualitative location data, these data should only be considered a secondary reference to be used in conjunction with the quantitative location.

What types of data are included? The actual information content of the data is a key part of a searchable database. This is probably the most difficult part of metadata indexing because it is the most subjective. For example, if a surface geological map with seismic shotpoint locations were indexed with keywords such as "geology" and "seismic," it might be retrieved incorrectly in a search for subsurface geology and seismic interpretations in an area.

The best way to make the indexing process successful is to predefine a lexicon of descriptive keywords that can be selected from drop-down lists during the indexing process (see *Data Validation, Editing, and Quality Control* and *Designing the User Interface* chapters). This eliminates the errors introduced during data entry by multiple workers and yields a much more consistent and accurate set of metadata. In some cases, there will be multiple types of data on regional reconnaissance documents. All appropriate data keywords should be indexed into the data type metadata.

What is the source of the information? An often-overlooked type of metadata is information about the source of regional and/or reconnaissance data. Much of the data used during this phase of the exploration-development cycle come from old maps, reports, and other documents that may not have an obvious source or owner. However, it is important at this point to collect as much information as possible regarding the original author and/or owner of the data, and any information about how the data were acquired (purchase, trade, asset acquisition, etc.).

Gathering metadata-indexing information is tedious and repetitive and is generally relegated to the least experienced staff member, an assistant with no geotechnical background, or a non-technical consulting firm. While accurate indexing requires the attention of the most experienced individual available, it is impractical to assume that the senior person can be devoted to such an "unrewarding" task. Therefore, one needs to develop an indexing system that incorporates the detailed business rules into the indexing system. Using an indexing system that offers a predefined set of indexing keywords provides the person doing this work with guidelines established by more experienced workers. In addition, viewing the database with some sort of spatial data browser (a GIS-based interface) will help identify major problems with the spatial indexing (see *Spatial Data* section in *Geotechnical Data* chapter).

Are there any special considerations? A large amount of reconnaissance data will include information derived from various sources. These sources include data that have been traded, purchased, or incorporated as the result of consolidations and mergers. Many of the original sources may no longer be viable corporate or individual entities. If information is available on any restrictions placed on the data that limit use, distribution, sale, or trade, this information must be included in the database. As stated above, the length of the exploration-development life cycle generally exceeds the length of time that any one individual works on the project. Knowing what data assets are available and what restrictions and/or limitations are placed on that data will be extremely important when the asset is sold or abandoned.

Reconnaissance phase data types and issues

The reconnaissance phase of the asset life cycle will include several unique data types that may be used only during this phase. Commonly, these data types do not fit neatly into most petroleum data management systems and may require somewhat customized solutions for effective data management.

Regional and remote sensing data. A wide variety of regional data types and remote sensing data products can be incorporated into a reconnaissance project, including (but not limited to) the gravity and magnetics data, surface mapping data, and imaging data.

The database must account for multiple vintages of surveys, which may incorporate marine, airborne, and/or surface *gravity and magnetics data*. In addition to all of the technical data related to the survey parameters, it is very important to determine the ownership, survey coordinates, coordinate projection system, and other key location metadata information (see *Geophysical Data* section in *Geotechnical Data* chapter).

Among the various types of *surface mapping data* that may be available will probably be geological maps, geographic information, cultural features, political boundaries, and facilities (pipelines, refineries, etc.). In many cases, known deposits of oil, gas, and minerals (existing fields, known seeps, etc.) may be mapped.

Imaging data can extend from the most mundane of high-altitude conventional aerial photography to sophisticated multispectral satellite imagery. Specialized georegistered satellite images, synthetic aperture radar (SAR), side-looking airborne radar (SLAR), and other similar remote-sensing data may also be present.

Direct hydrocarbon indicators (DHI). Direct evidence of oil, gas, or other related economic mineral deposits is extremely valuable information for the explorationist, especially in the early phases of a reconnaissance project. In many cases, this information may be included or incorporated into other types of displays and may not be intuitively obvious to anyone but an experienced worker. Following are some of the types of DHIs to be aware of from a data management standpoint.

The presence of *oil, gas, and/or tar seeps* may be indicated on surface geological maps, documented in regional reports, or described in local cultural documents or maps (some may be historical sites or places known only to local residents). To be useful, the location of these deposits must be recorded accurately in the database with the appropriate coordinate system, map projection information, and other pertinent data.

The location of *tar deposits* may be incorporated into surface geological maps or described in reports on economic mineral deposits. While an asphalt mine may not be of immediate interest to someone exploring for subsurface oil and gas, the presence and location of these deposits need to be included accurately in the database.

Surface geochemical surveys are unique and can include soil and plant geochemistry, vegetation surveys, airborne surveys, and offshore (sub sea) studies. Obviously, conventional information regarding the source of the data, overall location of the survey area, and pertinent information about the survey data in general are the basic metadata needed. Specific information about the survey results, such as the various hydrocarbon components, concentrations, and other detailed geochemical information, will depend largely on the survey itself. Ideally, all data of this type should be collected and reviewed before beginning indexing and data capture.

The most valuable information for a reconnaissance study, but in many cases the least common in new or underexplored regions, is *subsurface data*. However, as information from drilling and seismic surveys is added to the database, it is helpful if the database has been constructed in such a way that the new data can be incorporated seamlessly with little or no retroactive modifications to historical data.

Many older wells may not have a large volume or wide variety of *well data* or borehole information other than drilling data, such as penetration rate, weight on bit, or bit rpm. Much of the early drilling data, including electrical, acoustic, or nuclear logs, are not readily available in digital format. Since drilling information and historical log data are not always suited for quantitative analysis and interpretation, it is many times more practical to scan, index, and store the data as documents for review by the geotechnical interpreters (see *Log and Borehole Data* section in *Geotechnical Data* chapter).

Many older wells will include *subsurface samples* such as cuttings sample descriptions and logs and in some instances core data and fluid samples from drillstem and/or production tests. Lacking sophisticated wireline logging tools, early drilling evaluation relied heavily on cuttings sample descriptions; these data should not be ignored or underutilized. As with historical wireline logs, however, these data are better stored as scanned images (with appropriate metadata indexing) rather than attempting to convert them to digital data.

Production and drillstem test data, including amounts and types of produced fluids, commonly provide a great deal of information about the presence of hydrocarbons in a region. As with most reconnaissance data, it is sometimes better to collect all related data of this type before determining how to store the information in the database.

Surface sample data. Most surface sample data are very specialized and are source driven. That is, the data storage method and format are determined by the content and source of the original data. Surface data are diverse and highly variable and can include rock petrology, petrography, geochemistry, and source rock studies. Much of the information is written, with or without photographs, and can best be handled as scanned documents in a document storage/retrieval database. The possible exception to this is source rock maturity and quality studies, which include a certain amount of quantitative data. Thermal maturity or basin analysis studies commonly include basin maturity or subsidence profiles.

Where available, the original data used to generate these plots should be stored in the database, and the subsidence profiles should be scanned, indexed, and stored in a document storage/retrieval database.

Data integration issues. During the reconnaissance phase of the exploration-development cycle, integrating the wide variety and types of data is the most significant challenge facing the geoscientist and the data manager. Some issues to consider follow.

Diverse data types are a big issue. The greatest consideration in data storage is the accurate and complete indexing and collection of metadata (the basic information about the data content). If a consistent indexing method is used, the data will be integrated much more easily and effectively.

In working with regional data, there will be a wide *variety of map scales, data scopes, and data content* coverage. Because of this variability, the use of a geographic or geospatial information system (GIS) browser should be seriously considered. A GIS browser interface allows easy integration of different scales, scopes, and vintages in the same spatial environment.

At the outset of a project, it is essential to have a *data inventory*—the basic what, where, and how much information about the data availability. Ideally, this fundamental data inventory can be generated directly from the database (once everything has been cataloged, indexed, and stored). As will be discussed in later sections, this data inventory is important during several phases of the exploration-development cycle.

Establishing fundamental business rules regarding the *source and ownership of data* should be done at the very beginning of a project. Knowing the origin and lineage of a particular data item not only establishes the validity and reliability of the information but also becomes critical in later phases of the exploration-development cycle.

Visualization and interpretation solutions. As indicated above, the use of a *GIS interface* for integration and visualization is a valuable tool. GIS provides a map centric solution to the data integration problem and allows integration of data from various sources at different scales, scopes, and vintages (see *Interface Design Options* in *Designing the User Interface* chapter). This approach requires a very consistent approach to collecting metadata, especially with regard to the spatial information (coordinates, projection systems, etc.).

Because of the heavily quantitative nature of reconnaissance data, it is probably more important to have accurate and complete indexing and coordinate data than qualitative information on the data content. Once the data can be spatially compared and viewed, it is easier to decide what is important and connected and what is extraneous. If all of the data are indexed, even if the actual documents are not scanned, the content can be retrieved and used much more easily.

A key element in the evaluation of a region, *basin models* determine if the basic elements of source, reservoir, trap, and seal are present in a configuration that favors the development of commercial accumulations of hydrocarbons. By its nature, much of the data associated with basin modeling are subjective, interpretive, and text focused. Therefore, a document management system is the best vehicle for storing, managing, and retrieving data content. As always, however, it is critical that accurate spatial (location) data and keyword indexing be completed at the time of document capture.

One component of basin modeling, thermal modeling, includes quantitative information on source rock maturity, migration timing, and subsidence profiling. While custom data models can be developed to store and manage this information, it may be more advantageous to use commercial solutions that are developed specifically for this information.

As the evaluation of an area progresses, prospective *trends, specific play types, and individual prospects* begin to emerge. At this point, information about these different components should be captured and included in an enterprise-wide database. A large organization will need this information (along with economic and reserves forecasts) for planning, budgetary, and strategic purposes.

Once specific prospects begin to develop and mature, there will be numerous *reserves/resource estimates* made and a wide range of economic evaluations and forecasts generated. The assumptions made for estimating reserves and the parameters and methods used to develop the economic forecasts must be recorded in the database. This ensures that the estimates are repeatable and that consistent parameters and methods are being used throughout the organization. Again, specialized databases may be more suitable for storing these data than custom-developed solutions.

EXPLORATION DRILLING PHASE

During the reconnaissance phase, one or more drilling prospects will be identified in areas that are prospective for hydrocarbons. As exploration wells are drilled in an area, data volumes will increase and some new data types will be added to the database. Many of the data types identified during the reconnaissance phase will be expanded and extended. If the basic principles of data management have been applied, much of the groundwork will have been done. Figure 3–3 illustrates the general importance of various data types during the exploration drilling phase of the exploration-development life cycle.

Exploration and Development Asset Life Cycle Matrix					
	Life Cycle Stages				
Data Type	Reconnaissance	Exploration Drilling	Delineation Drilling	Exploitation	Abandonment
Remote Sensing					
Maps					
Text					
Well Data					
Production Data					
Data Volume					
Data Management					
Visualization					
Key: Data Management Importance	Very High	High	Moderate	Low	

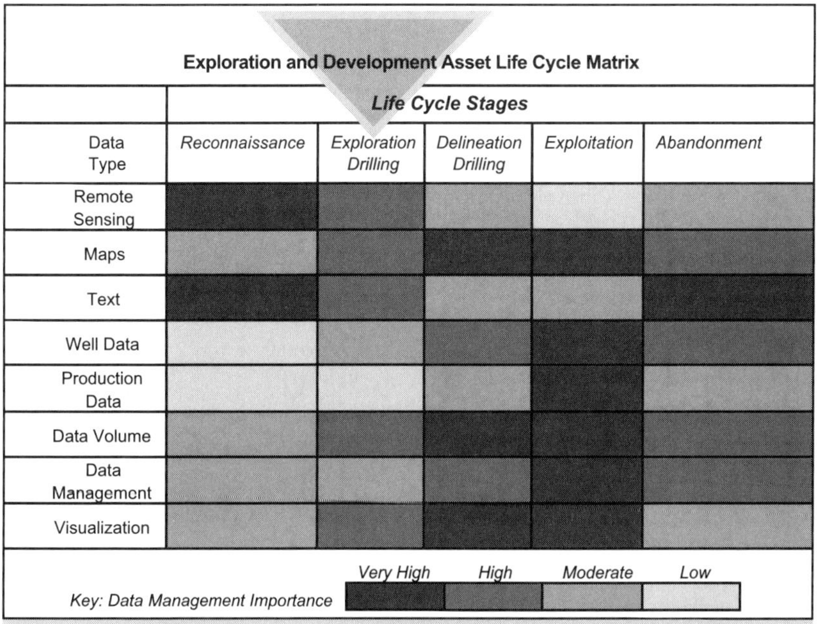

Fig. 3–3 Exploration Drilling Phase of the Exploration-Development Life Cycle. The relative importance of major geotechnical data types is highlighted.

Acquisition and interpretation of exploration drilling data

Although a certain amount of drilling data may be available during the reconnaissance phase, the exploration drilling phase will probably add large volumes of new data.

Typically, exploration wells are designed to acquire a wider variety and larger volume of data than later phases where cost containment becomes paramount. While this presents unique challenges for the data manager in defining data types and indexes, the resulting database framework will accommodate virtually all future drilling data types. Some of the new data types that will be introduced at this phase of the exploration-development cycle are discussed as follows.

Drilling operations generate large volumes of both qualitative and quantitative data. Much of this information is text based and can be accommodated in an existing document management system (with appropriate changes to the metadata indexes to include the new data types). The quantitative data are very specialized and may be better managed with a specialized operations data management system such as the DIMS® product from Landmark Graphics. The most critical drilling operations data to the geotechnical user are the physical well location and borehole geometry data. Accurate surface location and elevation data are essential for positioning all subsequent subsurface information (see *Coordinate Data and Depth-Related Data* in *Geotechnical Data* chapter). Borehole geometry data, in the form of directional survey information, provide the only means of accurately locating subsurface features (see *Directional Survey Data* in *Geotechnical Data* chapter).

If the log data management system has been properly addressed and designed in the reconnaissance phase, the storage of *open-hole well logging data* at this point will involve some modifications for data types, log types, and other minor modifications (see *Log and Borehole Data* in *Geotechnical Data* chapter).

Many of the database components designed during the reconnaissance phase for managing surface samples can be adapted (with minor modifications) for *subsurface samples*. Since the data acquired in this phase of the exploration-development cycle will include sample logs, drilling logs, conventional and sidewall cores, drill cuttings, etc., there will be a need to index these data by well rather than by geographic location.

As exploration wells are drilled, abundant *testing and sampling data* will be acquired. This will include drillstem testing (DST), repeat formation testing (RFT), multiple formation testing (MFT), and in some cases, testing through perforations.

Any *fluid samples* recovered during the testing process will probably be analyzed, so the analytical data need to be added to the database. Again, some of this information will be quantitative in nature (and can be stored as numerical data in the database), but a large amount will be qualitative and textual, which is better handled in a document management system.

Shift from surface to well-centric subsurface data

Fundamentally, the exploration drilling phase shifts the focus of data management from predominantly surface information to subsurface information. The difference between the spatial, coordinate-based information of the reconnaissance phase and the well centric subsurface data of the exploration phase will be carried through all subsequent phases of the exploration-development cycle. Therefore, the planning and execution of data management issues are very critical. Following are some of the most critical points to consider during this process.

As is discussed in subsequent chapters, *clearly defined well nomenclature* is essential for the effective management of predominantly well centric data (see *Stratigraphic Tops, Zones, and Markers* section in *Geotechnical Data* chapter). Well naming standards must allow for flexibility, expansion, or extension and should incorporate government or other existing nomenclature wherever possible. If the nomenclature business rules are defined early and adhered to throughout the project, retroactive modifications at later stages (which can sometimes be very impractical, if not impossible) will not be necessary.

For the same reasons discussed under well nomenclature, *subsurface (stratigraphic) nomenclature* should be defined at this stage. Because of the extreme variability of stratigraphic nomenclature, it is sometimes helpful to convene a committee of senior geologists to establish guidelines and business rules for stratigraphic nomenclature (with the data manager to moderate and serve as a reality check during this process).

Before the extensive use of intentional directional drilling and multilateral wellbores, it was conventional practice to associate all subsurface data with the well name and well surface location. It is now more likely that there will be multiple physical wellbores associated with any given well and surface location, so the well/borehole nomenclature should *associate information with the physical borehole* rather than the well itself (see *Log and Borehole Data* section in *Geotechnical Data* chapter).

Geophysical data acquisition and interpretation

Although some geophysical data are available during the reconnaissance phase, large volumes of data will be added just prior to and during the active exploration drilling phase of a project (see *Time-Related Data* section in *Geotechnical Data* chapter). Furthermore, old and new data will be added through trades, purchases, acquisitions/mergers, and during government-sponsored exploration bid processes. Considerations must be made for managing this new information.

In most areas, there will *multiple generations of geophysical data*—a wide variety of data types (3-D, and sometimes 4-D), acquisition methods, data densities, and processing methods.

One of the most difficult parts of the geophysical data management process is handling *multiple generations of interpretations and processing methodology*. Data management applications that are specifically designed to manage geophysical data are probably more suitable than attempting to create a system from the ground up. Some consideration should be given to combining a text or document management system with the application used to manage the actual data as a means of dealing with the complex mix of documentation and huge volumes of quantitative data.

Borehole geophysical data acquired from borehole measurements include vertical seismic profiles (VSP), borehole gravimeter data, and other specialized data. These data represent a special challenge, as all handling of conventional seismic data must be considered in conjunction with the well or borehole models used for well-based subsurface data.

As discussed earlier, there will be multiple geophysical data sets in most areas that are obtained from *various sources or owners*. The ownership and origin of the data sets should be clearly and carefully documented at the earliest opportunity in the project. Particular attention should be paid to any contractual, legal, or governmental restrictions and limitations placed on the data because this will become a key issue in later phases (during asset disposals, mergers, etc.). Most of this information will be available early in the project's life cycle.

Virtually all other data types (with the exception of satellite imagery and subsurface image logs) require relatively small amounts of *storage space* from a data management standpoint (see *Data Storage Requirements* section in *Planning Database Projects* chapter). Geophysical data, especially newer 3-D and 4-D surveys, require huge volumes of storage space to manage the original data volumes, multiple processed volumes, and multiple working project sets and individual interpreted volumes. Planning and accounting for this data volume (and the storage/access speed requirements) require close coordination between the data manager, acquisition geophysicists, processors, and interpreters. In most cases, it is better to account for more space than you expect to use.

Backward integration

Results from the early part of the exploration drilling phase will undoubtedly change or modify many of the concepts developed during the reconnaissance phase. These reinterpretations may change earlier defined trends, leads, and prospects and will have both positive and negative impact on the entire project. The use of integration and visualization tools will provide an effective environment for comparing previous concepts and data with newly acquired drilling data. Software suites that provide a multidisciplinary interpretation, integration, and visualization environment are the best choice at this point (see *Database Applications* section in *Planning Database Projects* chapter).

It is essential that any application be directly linked to a formal database so all data are immediately accessible and changes to the data are reflected back to the database in real time. In the same way, visualization tools should be linked directly to a compatible database/data management system, especially if the application provides functionality for data manipulation and editing.

Forward modeling

As new exploration wells are drilled, the hope is that new plays and prospects will be generated. A data management system needs to include some means of maintaining an inventory of prospects, leads, and drilling locations. If this capability was not added during the latter phases of the reconnaissance phase, it should be done early in the exploration drilling phase. If possible, the prospect, lead, and drilling location database should be linked (through the database) to a GIS visualization browser. This also provides a convenient environment for comparing data, interpretations, and planning aspects. At this point, numerous preliminary economic analyses will have been completed and should be included in the database (even if only in a document management system).

During this and all subsequent phases of the exploration-development cycle, it is very important to maintain continuously updated economic appraisals and reserves assessments. In this way, the economic performance, reserves, and potential of an area can be more effectively managed through a corporate, multiple asset portfolio management approach.

FIELD DELINEATION PHASE

The distinction between exploration drilling and delineation drilling is subtle and indistinct from a data management standpoint. Most of the same data types, data volumes, and problems found in the exploration phase are present during the delineation phase. Figure 3–4 illustrates the general importance of various data types during the field delineation phase of the exploration-development life cycle.

Exploration and Development Asset Life Cycle Matrix					
Life Cycle Stages					
Data Type	Reconnaissance	Exploration Drilling	Delineation Drilling	Exploitation	Abandonment
Remote Sensing					
Maps					
Text					
Well Data					
Production Data					
Data Volume					
Data Management					
Visualization					

Key: Data Management Importance — Very High, High, Moderate, Low

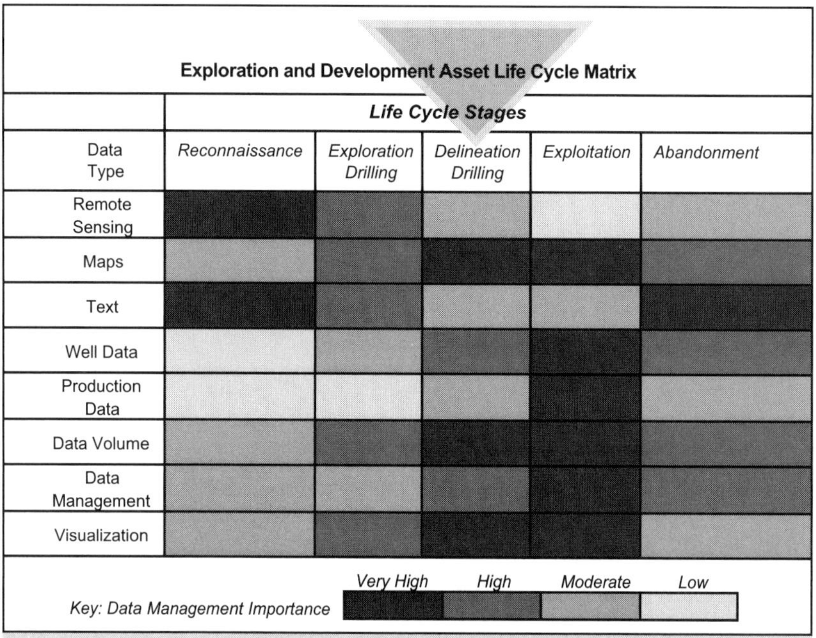

Fig. 3–4 Delineation Drilling Phase of the Exploration-Development Life Cycle. The relative importance of major geotechnical data types is highlighted.

Impact on data management

The focus of data management during delineation drilling shifts more toward drilling operations and related activities. At this time, more effort will be spent on field delineation activities and less on the big-picture issues of the first two phases of the exploration-development cycle. Although prospects and leads continue to be updated, there will be a gradual need for creating and maintaining individual field models. At this time, there will be a rapid increase in the pace of data acquisition, which means there will be a critical need for rapid and accurate integration of new data as well as the updates and revisions to existing data and interpretations. This again illustrates the fundamental importance of developing and implementing standards and business rules early in the project life cycle (see *Importance of Standardization* section in *Planning Database Projects* chapter).

The acquisition-integration-interpretation data life cycle is also shortened significantly. Thus, the database must be accessible and integrated completely by all disciplines because there will be little or no time for experimentation, learning new systems, or significantly modifying existing systems.

Shift to production and engineering data

If the field delineation phase is even moderately successful and the process confirms the commercial presence of hydrocarbons, well testing and production data will be available, although in relatively small, manageable amounts. Production data need to be integrated into the data management system so that well and reservoir performance can be compared to reserves modeling and economic forecasts. This, in turn, leads to inevitable modifications to reserves forecasts and economic analyses.

As more production is generated, there will be the need to start incorporating surface facilities information into the database, although there will only be a limited need for tight integration between production data and geotechnical data. However, the surface facilities can be integrated into the GIS visualization browser interface to plan future facilities and select drill sites. Certain specialized engineering data are generated at this point, but the management of this information is beyond the scope of this book.

Data volume impacts

During the field delineation phase, there is a rapid increase in both the volume and type or scope of various data types. Open- and cased-hole logging data are generated in increasing amounts. Geophysical data, in the form of refined 3-D and possibly 4-D surveys in the active parts of the field, need to be managed. With the increase in data volumes, there is an increasing need to develop data normalization procedures and standards (see *Data Normalization* chapter). Multiwell reservoir models can be developed and should be integrated into the overall data management system. Field-wide standards should be developed where appropriate, especially with regard to field-specific data types, stratigraphy, and drilling procedures.

Development Phase

The development phase, defined here as infill drilling and primary production for a defined field, requires a shift in focus to economies of scale and an overall reduction in data acquisition programs. Data collection generally is limited to that which directly impacts the bottom line in terms of reservoir development. Much less regional information is purchased during this period because most of the activity focuses on detailed field operations work. However, there may be several opportunities to capitalize on the value of regional data in areas not being actively developed. If unexplored or underdeveloped assets have been identified as low priority or unattractive to the company, costs associated with the acquisition of the data associated with those assets can be recovered through trades and/or data sales (see *Data Transfer Issues* section in *Planning Database Projects* chapter).

Data interpretation

During the development phase, geotechnical data become much more multiwell and multidiscipline oriented. Integrated multidiscipline studies draw on existing databases for interpretive raw data. The results of these studies, however, generally are written documentation, maps, and graphical results that can be handled using document management applications.

Engineering and simulation *modeling studies* become more important. The data used in these studies can be exported and transferred using industry-standard file formats such as RESCUE or ASCII flat files.

Economic modeling and *reserves forecasting* continue to be very important during this phase. It is important to continue to maintain accurate, multicase reserves studies that model the downside, most likely, and upside reserves and economics for the asset. Economic forecasts should be adjusted as product price scenarios change to ensure that the most current analyses are available at all times.

The *exit strategy* for the asset should be continually updated with current reserves, production forecasts, and economic analyses. From a data management standpoint, a current and complete inventory of data assets should be maintained as part of the exit strategy. This way, if a rapid decision is made to dispose of the asset, no time will be lost assembling and organizing the data for review by prospective buyers.

EXPLOITATION PHASE

During the exploitation phase of the exploration-development cycle, the driving forces become critically bottom-line oriented. This period is considered by many to be the harvest phase of an asset. Data acquisition during this phase is limited strictly to that which is necessary for continuing operations. Any new data acquisitions must be demonstrated to be necessary and cost-effective because any expenditure negatively impacts the low-margin opportunities. Data mining and statistical analyses of existing databases are instrumental in determining the need for, or providing justification for, additional data acquisition. Most additions to the database at this time are interpretations based on existing data or new interpretations using existing data. Figure 3–5 illustrates the general importance of various data types during the exploitation phase of the exploration-development life cycle.

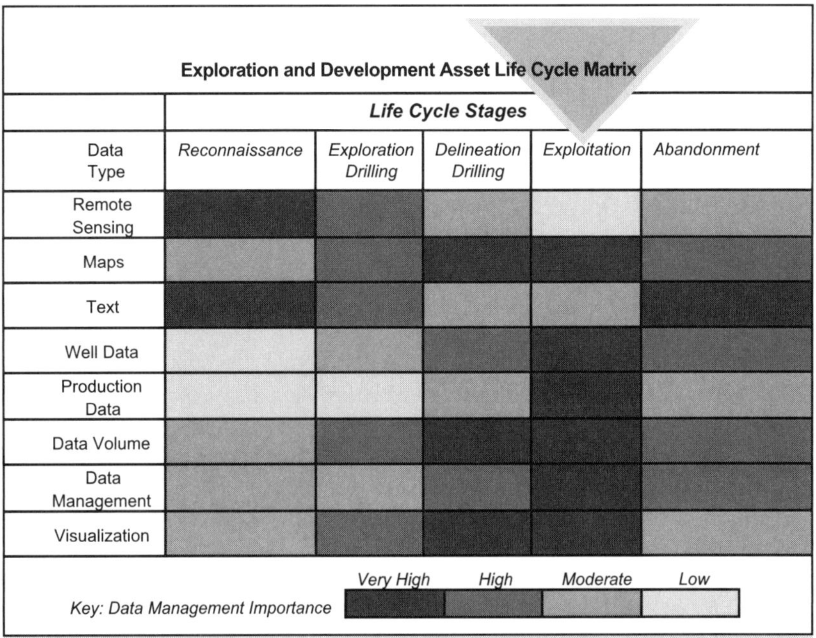

Fig. 3–5 Exploitation Phase of the Exploration – Development Life Cycle. The relative importance of major geotechnical data types is highlighted.

Virtually all activities during this phase require the rapid and accurate integration of huge volumes of data, making inexpensive and effective data access mandatory (see *Data Utilization Requirements* section in *Planning Database Projects* chapter). Primary activities during this period probably focus on enhanced recovery projects, including waterfloods, gas injection, steamfloods, miscible floods, and other special enhanced recovery projects. These activities are concurrent with ongoing workover and redevelopment work as well as overlooked production evaluations.

Hardware requirements

One of the major changes during this period should include a number of hardware and software modifications (see *Scalability and Portability Considerations* section in *Data Types and Formats* chapter). High-end hardware can be reallocated to less mature assets or properties that require more computational horsepower. Data management solutions therefore must be scalable to run on various platforms, operating systems, and network configurations. As staffing levels are reduced or reallocated, support limitations require that simple, bulletproof applications be preferred over full-featured, high-end data management and delivery systems. Inexpensive, cost-effective solutions become more practical because new data additions are minimal during this phase.

Software considerations

Software requirements during the exploitation phase also shift dramatically (see *Database Applications* section in *Planning Database Projects* chapter). Licenses for full-featured, high-powered interpretation applications may be reallocated to less mature or more complex assets requiring more sophisticated or complex applications. Concurrent with the hardware and software reallocations is a shift of interpretive manpower. One consideration of the exploitation phase is that simplified, custom-developed proprietary applications may be developed, tailored to the specific needs of the asset. These disposable solutions may greatly simplify the data management and delivery issues for the asset and provide an additional opportunity for cost containment and reduction (see *Customizing Commercial Products* section in *Designing the Database* chapter). The support and maintenance of these applications must be carefully considered in light of labor availability and costs.

ABANDONMENT AND REMEDIATION PHASE

During the asset abandonment and remediation phase, documentation again becomes one of the most critical data management issues. Extensive documentation and communications with regulatory and government agencies are usually required during the asset abandonment process. Much of this documentation is required to ensure and demonstrate compliance with internal health, safety, and environment (HSE) personnel and external regulatory programs. If the asset has remaining potential, documentation during this phase will be required by potential buyers, operators, or contractors as part of the asset custody transfer. If portions of the abandonment or remediation process are to be handled by a contractor or consulting group, this documentation will be crucial for the smooth transfer of control for the asset. Figure 3–6 illustrates the general importance of various data types during the abandonment and remediation phase of the exploration-development life cycle.

Exploration and Development Asset Life Cycle Matrix					
	Life Cycle Stages				
Data Type	*Reconnaissance*	*Exploration Drilling*	*Delineation Drilling*	*Exploitation*	*Abandonment*
Remote Sensing					
Maps					
Text					
Well Data					
Production Data					
Data Volume					
Data Management					
Visualization					

Key: Data Management Importance — Very High | High | Moderate | Low

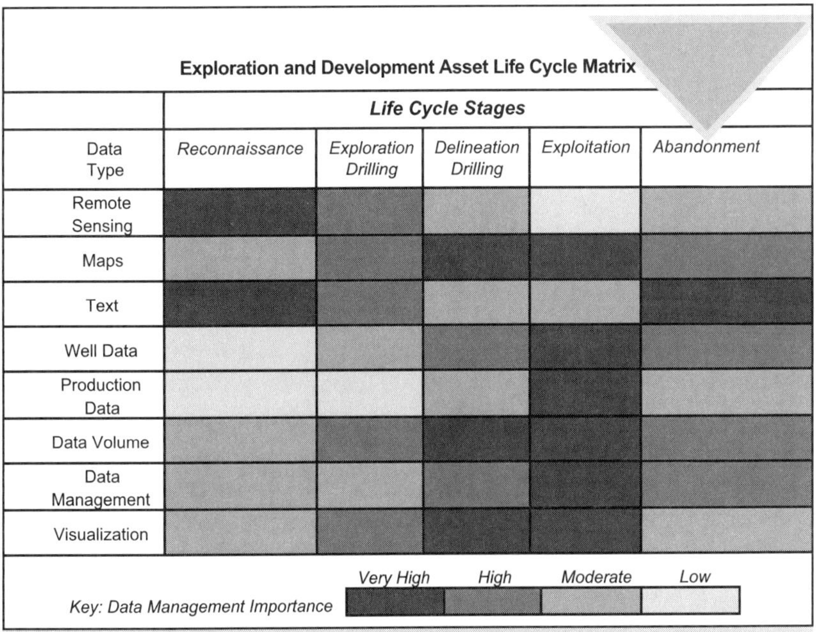

Fig. 3–6 Abandonment and Remediation Phase of the Exploration-Development Life Cycle. The relative importance of major geotechnical data types is highlighted.

At the end of the exploration-development life cycle, a great deal of experience and information has been collected about the region, trend, and developed fields. The documentation of this wealth of information is in the form of drilling post-appraisals, field studies, and case histories. As in the early stages of the cycle, these studies, reports, and other interpretations form the basis for other exploration or development opportunities. The same data management solutions, in the form of document management applications, are used extensively at this point.

Data storage considerations

At the end of the life cycle of an asset, cost, time, and labor limitations generally limit how much attention will be devoted to data management and documentation. As information is prepared for long-term archival storage, new concerns about media type, data formats, and maintenance schedules must be considered (see *Planning Database Projects* chapter). Again, document management is an integral part of the process and forms the framework for hard or raw, quantitative digital data. Despite the personnel limitations during this phase, accurate and complete indexing is essential for the successful future use of the archived data.

SPECIAL CASES: ACQUISITIONS AND DISPOSALS

The acquisition and disposal of assets are normally continuous and simultaneous processes in most large organizations. Both can occur at any point in the life cycle of any producing or nonproducing property. Data management staff and geotechnical professionals must be prepared at any point to deliver the information needed to evaluate a potential acquisition or to assist in the disposal of an existing asset.

Acquisitions

During the evaluation of potential acquisitions, data management is an integral task for the acquisition evaluation team. The data manager provides support for the team by preparing background data and performing data reductions, summaries, and analyses (see *Role of the Database Manager* section in *Planning Database Projects* chapter). The data manager can improve the efficiency of the data loading and interpretation process by standardizing data formats and ensuring that all required data are available to the interpreters (see *Data Reformatting* chapter).

Much of this work can be done in advance of the actual interpretation work, especially if the format standards are clearly defined for each application in use. If these standard format requests can be sent to the organization or company offering the asset or conducting the data review, the data can be preformatted in many cases to meet the needs of the interpretive applications. Of course, there will be instances where the data are provided in a specific format with no options for customization. If the data manager is aware of these problems ahead of time, appropriate format conversion plans can be planned and tested in advance.

Disposals

One of the most important parts of an asset disposal is preparing the data packages that will be reviewed and used by prospective buyers. These data packages need to be carefully filtered to include only the data directly related to the disposal to avoid unnecessary disclosure of any proprietary information that is not relevant to the sale. As discussed earlier, the data model should be planned and constructed early in the project life cycle to allow indexing of all geotechnical information by multiple levels of scale and scope (zone, well, pool, asset, field, trend, region, etc.). This will give the data manager more control over the information included (or excluded) from a data package used by outside evaluators.

Data formats and delivery methods

To facilitate the utilization of data packages by outside evaluators, several factors should be considered.

Whenever possible, use *industry-standard data formats* when preparing data packages for outside review (see *Data Format Standards* in *Summary* chapter). Most commercial interpretation suites have the ability to import and use a variety of standard formats. If proprietary formats are the only available option, some provisions for format conversion may have to be provided.

In all cases, geographic location information should be maintained in an industry-standard format, and all relevant *map projection constants* should be documented and available.

If not already part of the data management solution, a *GIS-based data browser* should be considered for the viewing, selection, and delivery of the asset data. The underlying database for the browser should be restricted to information related to the sale area. This permits the evaluation team to quickly review available data without compromising proprietary information. If possible, export functions should be incorporated into the browser to allow simple and direct export of the selected data in the format appropriate for the interpretive application.

Summary

As we move from this general overview of the interrelation between data management and the exploration-development life cycle, several overriding themes should be considered at each phase of the cycle.

Plan data management needs early in the project life cycle. Standards should be defined early in the life of the project, implemented quickly, and maintained throughout the life of the asset. Although there will be some changes and modifications necessary along the way, the basic standards will be preserved. To this end, there are several proven data standards solutions readily available; in most cases, a proprietary set of standards should not be developed. In all cases, it is easier to adapt the geotechnical data to fit a set of defined standards than to create standards from scratch, or on the fly. Custom extensions to existing standards are possible but should be limited to those dictated by operational needs.

Avoid customized, proprietary solutions. It is always very easy (and sometimes very necessary) to develop quick, custom solutions to various data management problems. Each of these solutions, however, comes with its own set of support and maintenance problems (see *Customizing Commercial Products* section in *Designing the Database* chapter). Migration of the solution required by operating system, hardware, and/or application changes is another hidden cost of this approach.

Most custom solutions are the brainchild of an in-house expert or guru and not the product of a commercial vendor. If these solutions are part of the critical path in data management, the dependence on that individual becomes a serious concern for the company. Any proprietary solutions should be considered disposable, written in standard, portable code with complete documentation, and should serve only as an interim solution until an application vendor provides a permanent, externally supported solution.

Allow for flexibility in the data management system. Any data management system adopted should utilize some sort of public or commercial data model rather than custom-built, proprietary data models. This will allow relatively easy migration of the actual data content to other systems or data models should that become necessary. A standard data model also provides a standard template for access by interpretive applications. As before, custom extensions and adaptations should be limited to those necessary to the operation for all of the mentioned support, maintenance, and compatibility issues (see also *Planning Database Projects* chapter).

Avoid change for the sake of change. Creating and maintaining a functioning geotechnical data management system is difficult enough without making constant changes that have not been thoroughly evaluated or considered. A well-defined set of business rules governing technology changes, application changes, data model changes, and format modifications should be put in place and adhered to by all affected parties in the organization, regardless of size. Evaluation templates and criteria should be developed and maintained to ensure that any changes in the data model, applications, hardware, operating system, or underlying data are compatible with the current system and future user needs.

In the chapters to follow, the details of project planning, data types, data management systems, editing, hardware, and software are discussed in more specific detail. The basic concepts, terminology, and relative importance of data mangement to the exploration-development process presented thus far will serve as a foundation for those discussions.

Planning Database Projects

Successful execution of a data management project is dependent on the level of planning that goes into it. Whether using a commercial database management system (DBMS), purchasing data and/or a DBMS from a vendor, or building a proprietary system, one must allow for a considerable amount of preproject planning. This chapter deals with the total project design, including setting the overall project objectives, selecting appropriate DBMS applications, and defining the project scope.

In addition to investigating hardware and software issues, we explore the role of the end user and how critical it is to involve the end users in the planning, development, and appraisal of the data management project. Other issues, such as inter-application data transfer and data storage problems, are introduced and discussed.

Defining Project Objectives

The most difficult and perhaps most important part of planning a data management project is defining the overall project objectives. If the objectives are well defined and clearly understood by management, users, and designer/programmers, the project is more likely to succeed. To make the planning process productive and orderly, it is always best to start by addressing some basic project objectives:

- Does the project represent a major data management effort that will affect the entire company and a significant number of end users, or will it be restricted to select technical disciplines and/or geographic areas?
- Is the project meant to solve a new set of data management requirements, or is it a replacement for a current or previous data management system?
- Must a significant amount of legacy data be migrated to the new system, only new data, or a combination of legacy data and new data?

Defining Data Management Objectives

Once the overall project objectives have been defined, the actual data management objectives should be addressed. This involves several steps, including defining the actual function and objectives of the database, the user interface, and the life expectancy of the total project.

Defining the function of the database

Several factors should be considered when defining the function of the database. At one end of the spectrum, the database may serve as the central corporate repository for all geotechnical data. At the opposite end, the database may only contain specific technical data for a particular discipline, user, or group. For example, consider if all geological and geophysical data will reside in the same central database or if each group of users will have separate and independent data stores. While it is sometimes easier to build (or buy) data-specific data management systems, in other cases it is preferable to have all the data in one place. To a certain extent, this is determined by user needs and the size of the database.

Defining the user interface

Possibly the most critical step in the data management system planning process is defining what will be needed in the user interface. User interface development is addressed more completely in the *Designing the User Interface* chapter, but some of the main planning considerations are as follows.

- Will it be necessary to provide a full-function, full-featured graphical user interface (GUI), including map-based GIS browsing capabilities? Map-based interfaces require more defined spatial coordinates and metadata.
- Will a simple, text-based interface fulfill the user requirements? While simple and easy to develop, this type of interface results in significant loss of flexibility and functionality.
- When defining these specifications, we must consider training, user sophistication, and the type of data being delivered. If there is a wide range of user experience, it may be necessary to develop both expert- and novice-format interfaces.

Life expectancy of the DBMS

Data management is, and will remain, a dynamic and changing environment. It is sometimes preferable to deliver a system to the user community that will provide a solution immediately, recognizing that this solution may need to be replaced or enhanced in the near future. If the need is immediate, the system developed must provide an immediate solution, even if the solution has a limited life expectancy. It is easier to upscale a system that users have embraced rather than delay development for months (or longer) while a more sophisticated system is developed. During the delay, the users will find and use alternative solutions for their data management problems, eliminating the need for the new system even before it is deployed.

Defining Specific End-User Needs

The most common failure in any data management effort is not understanding exactly what the end user needs—not to be confused with what the end user wants, which may be impossible to deliver. Often, database managers with an IT background or programmers will attempt to deliver a product that is very suitable for a generic accounting database application but is unacceptable for geotechnical applications. These are two different problems: the user's inability to define and/or articulate actual needs and the implementation group's lack of understanding the unique aspects of a geotechnical DBMS.

End-users are, in fact, the customer or client for the data management product. Unfortunately, they are seldom consulted when a new data management application is introduced. This part of the planning process is critical for the success of the project. After all, if the customer (in this case, the end user) does not accept the product, there is no point in developing it in the first place.

Case history: scaled application development

In this case history, an international company had already selected a commercially available relational database management system (RDBMS) and was ready to start loading data for the users. Unfortunately, the team of information management specialists had planned the project with little or no input from the actual users. The original project schedule had called for a two-year period of data loading, quality control, and training before the actual data were provided to the users. Making matters worse, the commercial product ran only on UNIX workstations (most users had desktop PC access and only shared workstation access), did not have an effective user interface, and provided limited data loading and export capabilities (only to that vendor's applications, in fact).

After a series of interviews with the end users (and discussions with those who would be involved in data loading, QC, and delivery), it became clear that the users were already disillusioned with data management projects and felt the current efforts would only take time away from their already busy schedules. The users needed quick solutions and really didn't care about the big picture.

The solution applied in this case was to develop a temporary interface using a PC-based 4GL DBMS (in this instance, Visual dBase®). This provided a simple development platform that linked to the main Oracle® database tables via open database connectivity (ODBC) links and allowed rapid development of data conversion, loading, computation, and export features. Once the system was fully functional and had been thoroughly tested by the users, a more robust enterprise solution was developed in parallel using PowerBuilder®. Most of the Visual dBase® code was directly usable in the new version, and development and testing of the new system were conducted without disrupting the existing system. Eventually, the older system was phased out.

Involving the user in the process

To whatever extent possible, the actual end users should be directly involved in the planning and implementation of any DBMS project. If this is impossible, a representative cross-section group or users familiar with the objectives should be involved. This involvement should take place at several points in the planning and implementation process, including a planning survey, individual user interviews, group feedback and planning sessions, and some sort of direct electronic feedback system.

Planning survey. The best way to start the development process is to conduct a survey of end users before starting the project. This serves two functions. First, the survey results can provide valuable information about the current workflow processes, current application use, and user expectations for a new system. Second, the survey results represent a data management and usage baseline that the new project can be compared with at a future point.

The planning survey should include all obvious questions about the user as well as specific questions regarding applications, workflow, individual preferences, etc. These questions should include the following.

- What is the basic information regarding the user's background, training, and experience level? Knowing what the user's previous experience with data management and the interpretive applications will help put his or her comments and suggestions in the proper context.

While a geologist's comment on the functionality of a geophysical application may not appear significant, the different perspectives offered by users with varied background can sometimes be more helpful than the opinions of so-called experts.

- What are the current technical interpretation applications that will be used on a regular basis? For each of these applications, what are the data requirements, how are new data loaded, and how are data exported from the application?
- How will data be used by the client (end user)? It is important to determine if the user is going to perform complex mathematical manipulations on the data or if the data will be used to plot spatial points on a map. The nature of the data usage will have a significant impact on the user interface and any applications that need to be developed for the import, reformatting, and/or data export.
- How satisfied are the users with current methods of data handling? Surprisingly, the results of this part of the survey may show that the users find the current methods of data handling meet their requirements. If this is the case, initiating a DBMS project may not only be unnecessary but could be counterproductive. Like the saying goes, "If it ain't broken, don't fix it." While this may confound the management or IT group that favors a new DBMS, at least everyone is made aware that the actual users are very content with the current way of doing things.
- Are there possible improvements to existing systems? If a data management system is already installed and in use, can this system be improved or enhanced to meet the user's needs, or is a total overhaul or replacement needed?

Individual user interviews. The most valuable information about the user's role in the data management project can be gathered during individual or small group interviews and work sessions. Talking to users, observing the actual workflow during real-time interpretation sessions, and discussing better ways to handle data are essential to understanding the workflow and dataflow processes that need to be considered. User interviews should be done in the least disruptive manner possible, and the participants should be selected based on their current work assignments, familiarity with application software, and overall industry experience. Most importantly, these users should be volunteers and not those selected by management as expendable for a project that doesn't appear to have immediate, tangible benefits to the bottom line of the operation. In many cases, the power users—those who work with the applications the most often and are the most familiar with the problems—can be selected from an application license usage log file.

Keeping the user in the loop

After initial planning surveys and interviews have been conducted, the user community often becomes disconnected with the project and loses interest. Without buy-in and acceptance from the users, a DBMS project will fail. Maintaining the momentum of the project and keeping the user interest level and involvement high can speed development and reduce the overall project effort.

Assess planning survey results. After the user surveys and interviews have been compiled and summarized, it is important (to both the users and management) to provide an initial feedback session. The results of the planning survey will be important as a benchmark for comparison surveys that will be conducted after the project has been rolled out or deployed to the users.

Establish *ad hoc* user committees or groups. Whenever complex technical issues need to be resolved, it is usually better to have those decisions made by a small group (emphasis on the word "small") of appropriate users. For example, when establishing stratigraphic nomenclature there will be almost as many opinions on how to do it right as there are end users. If a "stratigraphic nomenclature committee" is established that is truly representative of the user community as a whole, agreements and compromises can be more easily reached, which in turn will generally be accepted and adopted by the entire user community. Participants in these groups should be selected based on experience, expertise, and technical credibility.

Provide regular progress updates. Most DBMS projects are extensive and nearly invisible (from a corporate standpoint). As such, people appreciate knowing that some sort of progress is being made on the project, especially if they have invested their time and effort into planning surveys, interviews, and user committees. Because the finished product is generally not something that can be deployed or introduced in stages, it is important to keep everyone informed about the actual progress to maintain interest and anticipation levels.

Conduct follow-up surveys. After completion of the project, or following a major deployment step, it is very helpful to resurvey the users. In many cases, a similar set of questions can be pulled from the planning survey so that a direct before-and-after comparison can be made.

Continuous improvement processes

Data management systems, like the underlying geotechnical data that reside in the database, are very dynamic. As such, it is as important to monitor user satisfaction and feedback as it is to monitor the quality of the data in the database.

The user as client. Too often, DBMS projects become an end to a means. Project staff members become so caught up in the technical issues of design, programming, and implementation that they lose sight of the ultimate purpose of the system—providing the end users with rapid access to accurate geotechnical data. The user is the ultimate client, and this is a situation where the customer is always right. End users may not always know what they want in a DBMS (that is the purpose of planning the project), but they usually know what they do not want. Listen to the users; remember that only when data are used can the information be transformed into knowledge.

Reporting bugs and enhancing the database. One of the best methods to maintain contact with users and provide them with a simple means of communicating problems and recommending improvements is through a feedback database. This approach should be incorporated into the database design whenever possible and practical. While there are several ways this can be accomplished, the basic concept is the same:

- Provide the user a simple, straightforward method to report problems either with the database application suite or with the underlying data.
- Provide the user a method of communicating suggestions for improving the database system, data delivery methods, or other improvements.

A later chapter (*Data Validation, Editing, and Quality Control*) discusses several methods of tracking, monitoring, and reporting problems with both the data and the DBMS.

Tailoring the Database to the Data

While it is not always possible or practical to select a DBMS before beginning the data management project, whenever possible the database itself should be tailored to the data. Attempting to "force fit" technical data into an unsuitable data management system will result in poor performance and user dissatisfaction.

Data storage requirements. When developing a new database system, it is sometimes difficult to estimate the data storage requirements. Even in existing systems, estimating future storage requirements is not always easy. As discussed elsewhere in this chapter, storage space is a function of data type, anticipated volume, and future expansions and additions to the database. In almost every case, initial estimates will be significantly lower than actual requirements. Some guidelines to consider when planning storage requirements are as follows.

Static database storage requirements are the easiest to deal with. Examining the data types that will be stored in the system, estimating the size of sample data populations, and considering the impact of future editing will make a reasonably accurate estimate possible.

Dynamic databases (active drilling and logging operations, new field acquisitions, expanding internationally, etc.) will be the most difficult to plan. In these situations, it is better to make a reasonable estimate and then design and build a system that can be expanded easily and inexpensively.

Data type greatly influences storage requirements. Large 3-D seismic volumes, high-resolution graphics and scanned images, and similar data types have a significant impact on storage requirements. If dealing with any of these large data types, it is better to overestimate the requirements and plan for more storage as with dynamic databases.

While most geotechnical data are relatively straightforward, there are *special types of data* that require more elaborate planning. Data types are discussed in more detail in subsequent chapters.

Data retrieval requirements. Early in the planning phase of a DBMS project, there must be considerable attention given to the process by which data can be searched, viewed, and retrieved from the database. To a certain extent, the amount of flexibility provided for retrievals is inversely proportional to the users' skills with data management tools. Several methods allow both novice and advanced users to access data at different levels, as discussed in the *Designing the User Interface* chapter.

Data export/transfer requirements. The single most important aspect of a database is the ability to extract information for use in analytical applications. If the DBMS cannot export the contents (data) in a format compatible with the user's application, the database will not be used. Various methods of data export and transfer are discussed in the chapters on *Data Reformatting* and *Data Loading and Input*.

Data utilization requirements. How the end user will use the data influences, to a certain extent, the type of system developed. If the primary goal of the system is to provide the end user with a dynamic, highly interactive, customizable product, a much more sophisticated approach is needed. On the other hand, if the system is only needed for routine data browsing and reporting, considerable time and money can be saved by developing a less sophisticated product. Key points to consider in terms of data utilization are as follows:

- How will the data be used by the end user? If the primary use is for data querying and browsing, then development should emphasize this aspect.
- Will printed reports be needed? If so, the user community should provide concrete examples of the types of reports needed. If a legacy system is in use, reporting examples from that system should be used, with notes from the users showing where enhancements and/or revisions are needed.
- Do the data need to be transferred in digital format to another application? If export functions are needed, the user community must supply examples and templates for all necessary receiving applications. Where possible, export formats should be standardized, but any custom format requirements need to be clearly specified during the planning phase of system development.

Data type consistency. During the planning phase, all data that will be managed by the system should be carefully examined and categorized into consistent data types. All data of one type (e.g., text data, dates, numeric formats) should be handled in the database in the same way. For example, all spatial coordinate data (e.g., latitude and longitude) should be stored as a decimal format with enough precision (decimal places) to accurately store the data for the most precise spatial data in the database. In this way, there is never a problem within the database when loading or exporting a particular data type (see *Geotechnical Data* chapter).

OTHER CONSIDERATIONS

Some of the other factors that should be considered during the project planning stage include the issues of support, maintenance, system functionality, and standardization.

Support

Developing a data management system requires a commitment to ongoing support for the users as well as for the system itself.

Complex systems require a much higher level of programming *support and maintenance*. If trained, qualified support personnel are not going to be available to maintain the system in the future, it is far preferable to develop a simple, easy-to-maintain system than a complex solution.

In every case, adoption of recognized *industry standards* for data models, protocols, operating systems, and data format should be encouraged. This will help future maintenance, modifications, and data migration to be done more easily, and it is more likely that qualified personnel will be available than with proprietary, nonstandard approaches. For example, every company has a few experts who can develop useful in-house solutions that serve the immediate needs of the user community. However, the survival of the system depends entirely on (in many cases) one individual. If that person leaves the company (which will eventually happen), it may be impossible to continue providing support and maintenance for the system.

No matter how sophisticated the user interface, there will come a time when additional options, features, or enhancements will be needed. In most cases, it is preferable to have some sort of *internal programming capabilities* in the system to solve these specialized, *ad hoc* requirements rather than make major changes to the entire system. Often, the programming or scripting options are proprietary and require that the user learn special programming methods or protocols unique to that system.

If possible, these programming capabilities (used to add additional functionality to the system) should also adhere to current industry standards. An example of this approach is the internal scripting language Avenue used by ESRI in the ArcView® GIS product. Although similar to many high-level programming languages, Avenue is specialized and requires additional programming experience. Future releases of ArcView® may incorporate the industry standard Visual BASIC® as the internal programming language. This will allow easier customization of the ArcView® product by a broader range of programmers and support personnel (as well as potential third-party product developers).

Maintenance is the least popular but most critical part of any application or system. Even simple database solutions require periodic maintenance; systems that are more complex require almost constant maintenance. Without regular maintenance, minor problems in a system soon become unmanageable and users will quickly abandon the system. No matter how a system is developed, maintenance considerations directly impact the life expectancy of the solution. Dedicated support personnel must be available to fill this role.

System documentation is the most often overlooked portion of any DBMS. In many cases, a rudimentary set of user instructions is put together quickly and complete system documentation and user instructions are never completed. ("There's never time to do it right, but always time to do it over.") From the beginning, the philosophy, basic concepts, design, implementation, and other features of the system should be clearly documented in a standard format.

User instructions should be written (or at least edited) by a user familiar with the product and reviewed by the developers for technical accuracy. When considering the purchase of commercial products, the user documentation and support documentation should be carefully inspected before considering a product demonstration. If the documentation is poorly written, disorganized, and confusing, it probably reflects a poorly developed, difficult-to-use, confusing product.

Hardware considerations

The DBMS cannot be developed without careful consideration of both hardware requirements and hardware availability. Most companies have a significant investment in computing hardware. Furthermore, the interpretive applications used by the user community may involve specific hardware with which they are already familiar. For example, if all interpretive applications that rely on data from the new DBMS run on PCs but the DBMS only runs on a UNIX-based workstation environment, there will be major problems involved in implementing the new system. Other considerations include where the data will reside (centralized or distributed), processing speed requirements, and operating system requirements.

Corporate vs. distributed databases. A common misconception, particularly among nontechnical users and managers, is that all corporate data can (and should) reside in a single, centralized database from which all users in all disciplines manage their data. Even in small, geographically restricted companies, this is generally not the most practical or advisable solution. For example, there are wide differences between financial, accounting, human resources, and geotechnical database requirements. This is not to say that the same underlying DBMS cannot be used for all the applications (see discussion later in this chapter). The consideration is more about where the data will be located and how the users will access it.

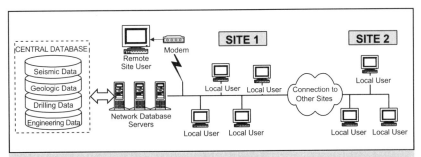

Fig. 4–1 Corporate Database with Local and Remote Access. In this example, a central database is accessed by multiple local users via a dedicated network, as well as remote users via dial-up or Internet connection.

Figure 4–1 shows a schematic representation of a centralized database system that is accessed by local network users, remote users, and other sites via a wide area network (WAN). A centralized database has several advantages:

- Lower administration, support, and maintenance costs
- Easier (scaled) development of common solutions
- The ability to develop and maintain quality assurance standards

Distributed databases, on the other hand, move many of the data management roles and responsibilities to departmental levels and/or widespread geographic locations. Advantages of distributed database management include the following:

- Allows development of custom solutions, both on a technical discipline basis as well by regional considerations
- Provides more flexibility in choosing vendors, support services, and application integration solutions

In most cases, the advantages of one approach become the disadvantages of the other. Only by examining the needs of the users in light of corporate objectives will one approach outweigh the other. There is no perfect, universal solution.

Processing speed and storage needs. Although the past decade has seen tremendous advances in technology, one trend remains clear: you always need more processing speed and larger amounts of storage space than originally anticipated. It is too easy to make shortcuts during the early stages of a data management project, most of which are regretted soon after the system is implemented. Obviously, it is important to plan and implement the fastest possible system with huge amounts of storage space, but it is more critical to examine the actual requirements, implement the most cost-effective solution that will meet those requirements, and design into the system the ability to expand in the future. Other considerations in this regard are as follows.

- Is the system designed mainly for storage and retrieval of data, or will there be additional data processing needs as well? As data processing needs increase, system-processing horsepower becomes more critical.
- Is the data storage static or dynamic? A completed rock catalog or paleontological database will require minimal future expansion

requirements, while a 3-D or 4-D geophysical database must anticipate ever-expanding storage and data management requirements.

Distributed databases. A distributed database is a complete, independent solution that provides data management and application functions on a stand-alone, or in some cases geographically local, computing environment. This type of solution is the most economical route for smaller companies or independent operators that have no significant data sharing needs. Figure 4–2 illustrates the simplest example of a distributed database, a single-user installation.

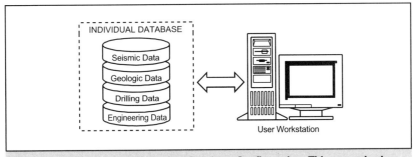

Fig. 4–2 Distributed or Independent Database Configuration. This example shows a single-user system that directly accesses a dedicated, stand-alone database.

Operating system requirements. Despite integration efforts and consumer marketing changes, there are still user applications running on PC, UNIX, and Mac platforms. These three groups have developed along technical lines for various reasons:

PC operating systems (Windows® in its various versions) have virtually taken over the workplace for administrative and personal technical applications, especially in the engineering environment. Because so many engineering applications require sophisticated analysis of relatively limited data at an individual level, this environment is generally preferred for petroleum engineering and reservoir engineering applications.

Workstation operating systems (UNIX, ULTRIX, Linux, etc.) are generally the platform of choice for geological, petrophysical, and geophysical applications. These applications require complex interpretive applications that use massive amounts of data and require sophisticated interactive graphical capabilities. Because of the limited graphical and processing support in early PCs, the workstation became the preferred environment. As PCs have become more powerful, the performance gap between workstations and PCs has narrowed. Clustered PC networks have successfully displaced even supercomputers in some instances. This trend will likely continue.

Apple Macintosh systems combine many of the best features of both the workstation and PC environments. For a variety of reasons, however, the Mac platform is still considered "consumer electronics" by many. Despite this unfortunate stereotype, the Mac excels at most graphical applications. Unfortunately, there are very limited native-Mac geotechnical interpretive applications. The graphics-handling capabilities of the Mac are unsurpassed, and support for this platform (if implemented) and the excellent applications that have been developed should not be ignored.

The users in these three camps in most cases have invested most of their professional careers in becoming proficient on whichever platform is most suitable for their needs. To develop a DBMS that is compliant to only a subset of the end users is, in many cases, counterproductive and can limit the successful implementation of the system.

While it is desirable to move all users to a single platform with a single interface and a limited suite of interpretive applications, this is not always the best solution. The best approach is to develop a DBMS that will serve most of the needs of most of the users while maintaining compatibility with other systems and applications. When possible, users should be encouraged to take advantage of a common operating system.

Recent advances and improvements in Web browser interfaces have demonstrated that multiple operating systems and platforms can be accessed via the platform-independent browser. This technology will continue to blur the divisions between workstations and PCs. Figure 4–3 illustrates a hypothetical system with multiple databases that are accessed by multiple applications from several different platforms and operating systems.

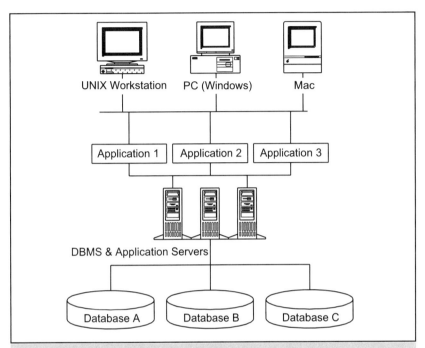

Fig. 4–3   Multiple Application, Multiple Database, Multiple Operating System Solution. This type of system will provide the highest degree of flexibility and functionality, but at a higher cost than a less-complex system.

THE DATABASE MANAGEMENT SYSTEM (DBMS)

The database management system software (DBMS) is the fundamental "heart" of the data management system. In some cases, it may not be possible to select the DBMS before the initiation of a data management project, as there may already be a system installed. The guidelines presented in this section should be considered regardless of the stage at which the DBMS is selected.

Selection criteria

The selection of a DBMS is made easier to a certain extent because there are a relatively small number of mature, proven DBMS applications available on the market. On a corporate scale, the most frequent choice is Oracle®. In the desktop world of PCs, Microsoft's Access® product is the most common application. That being said, these are not the only products available, and both products have their limitations and disadvantages. This section discusses the issues that must be addressed when assessing a DBMS product. Additional resources can be found in Appendix A.

Suitability to project objectives. Once the project objectives have been defined (see the preceding section), the functionality of the DBMS can be compared with those objectives. If the primary focus of the project is to develop a large, cross-platform corporate DBMS, a product like MS-Access® is not suitable. For a small, nonnetworked, individual database, an expensive, powerful, enterprise-scale product like Oracle® would be overkill.

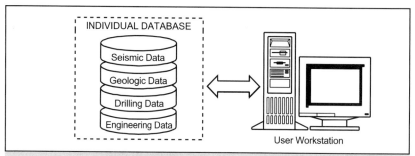

Fig. 4-4 Stand-alone Database-Application System. An independent solution like this may be practical for an independent or small company, but may not be suitable for a larger company with multiple geographical locations.

Expandability and upgrade options. If the project objectives are for a relatively static system where future changes and modifications are probably unnecessary (or for a system with a short life expectancy), the issues of expandability and upgrades are insignificant. Otherwise, it is very important to look carefully at how easily the DBMS can be expanded, enhanced, and upgraded. In the case of Oracle®, the ability to expand the system is only limited by storage resources and processing speed. For a purely PC-based system, more attention needs to be given to these problems. Because of the inherent limitations in the PC platform, there are fundamental limits to how far a system can be expanded. Obviously, with advances in networking and multiprocessor technology, storage space and processor speed will only be limitations for a limited time.

Technical considerations

The technical issues to be considered in selecting a DBMS in most cases require the assistance and recommendation of the IT staff who will ultimately support and maintain the system as the database administrator (DBA). These decisions should never be made without the approval of these individuals.

The main issues to be discussed with the software vendor and technical support groups should include the following.

- Will this be a single or multiple operating system or platform? The ability of a DBMS to run on multiple platforms and under multiple operating systems provides a much higher degree of flexibility and future options. If the DBMS chosen will only run on one platform or one operating system, future options are severely limited.
- Will this be a multiple or individual user environment? Multiple system users (which will comprise all but the most rudimentary data management systems) require networking capabilities (see below), shared access protocols, and multiuser support.

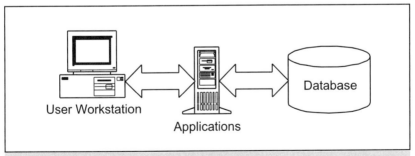

Fig. 4–5 Single-User Data Management Organization. In most single-user installations, the user runs interpretive applications on a PC or workstation. The actual software may reside on a server (as in this example), which also accesses the database.

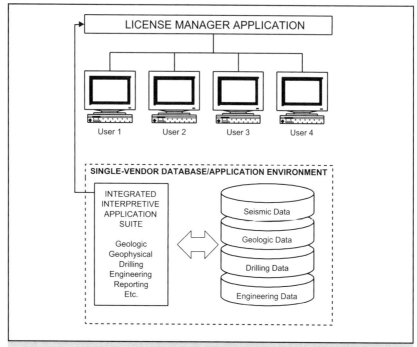

Fig. 4–6 Integrated Multiuser, Muliple-Application Data Management Organization. In this example, a single-vendor solution is provided which incorporates an integrated suite of interpretive applications that all access a common set of tables in a single database system. Most installations of this type use workstations for application access.

- Are there hardware compatibility problems? Does the DBMS run on the hardware that is currently installed or will there be additional issues related to installing and testing additional hardware? Will this new hardware be compatible with the systems currently in use?
- Will there be future compatibility problems? If the DMBS has a proprietary data storage structure or does not allow open connectivity to other systems, there could be problems in the future if the data need to be migrated to another system.
- Will the DBMS provide the speed and performance required by the users and volume of data to be handled? While desktop systems like Access® are suitable for smaller database implementations, there are upper limits to the number of records that can be handled, and performance tends to degrade as larger data volumes and users are added.

- Is the DBMS, in its off-the-shelf configuration, able to handle the basic requirements of the database project? Often a DBMS application vendor promises enhancements to a product to match the client's requirements. Based on the project objectives, the fundamental capabilities should be the minimum acceptable specifications.
- Does the DBMS have a built-in programming language and the ability to create user-defined functions? Without those capabilities, customization of the product (which is always necessary at some point) will be impossible or impractical. If a built-in programming or scripting language is available, it should be a recognized industry standard if possible.
- What capabilities does the DBMS have for exporting data or directly linking the database to defined applications? Part of this issue involves the ultimate delivery of the data to the end user's interpretive application. Export capabilities also provide an insurance policy against system obsolescence (i.e., data can be exported and migrated to another DBMS if needed in the future).
- What are the capabilities of the system when running on a network? Does the DBMS support client/server operations over a local-area network (LAN) and wide-area network (WAN) network architecture? How many users can run concurrently without adversely affecting performance?

Nontechnical considerations

The foregoing hardware and software considerations are very important in selecting a data management application (and its attendant operating environment) from a technical standpoint. Nontechnical factors also must be included in the overall decision-making process.

Hardware and software support. Although most modern software applications are quite stable and reliable, the increasing levels of functionality have led to more complicated installation procedures and complex support issues. Most software vendors provide some level of basic support but generally under a separate contract issue with separate pricing. Special care should be taken when negotiating software contacts to ensure that the appropriate level of support is included. If the application is relatively simple and reliable, it is possible to get by with less direct support. On the other hand, complex enterprise-level applications may require dedicated, onsite support. Support costs can quickly surpass the initial software expense.

When considering application support, it is important to look at the level of commitment and training of the vendor's support team. If the support group includes experienced staff with at least some level of geoscience training and experience, they can provide faster and more accurate response to support questions. It is reasonable and prudent to request details on the technical background of the staff members providing direct product support.

The issue of application support becomes a major issue in foreign locations. Because of the high cost of maintaining experienced staff in remote international locations, many software vendors utilize less-experienced local support personnel. In some cases, vendors provide no local or onsite support. If this is the case, you must rely on telephone support or (worse) e-mail support for problem resolution. Resolving complex technical problems by e-mail is generally less than acceptable. While telephone support centers provide rapid access to (sometimes) experienced support personnel, it is also necessary to consider time and work schedule differences.

Hardware support is a factor that most geoscientists would prefer not to consider. Without computer and network support, however, even the most sophisticated application becomes useless. Most often, existing hardware support is already provided, but new hardware may require additional support staff or maintenance contracts.

Case history: multiple time zone vendor support. This case history illustrates an extreme example of the problems associated with obtaining technical support from a software vendor that is not located in the same city (or time zone) as the client's installation. In this example, the application is installed in the client's office in the Arabian Gulf region. Although sales support is conveniently located locally, the main technical support office is located in Houston, Texas. This juxtaposition provides very little overlap in time for direct telephone support, even with extended hours at both ends. Furthermore, the international office observes a local work schedule where the "weekend" is Thursday and Friday, while the support office follows the traditional North American Saturday-Sunday weekend schedule.

With this in mind, consider the situation when a technical problem arises early in the week (Saturday morning) at the international client office. Although the problem is reported via e-mail, the technical support office (Houston) is closed until Monday morning. At that point, the support staff reviews the problem but cannot contact the client office until late Tuesday afternoon (client time). If the problem is not resolved then, another day might pass, at which time it is late Wednesday afternoon and the start of another weekend at the client office. A simple support issue, which might otherwise be resolved in a single day domestically, can take 7-10 days to resolve with remote technical support.

In this situation, two methods might be used to improve and speed the resolution of technical support issues. First, establishing set times for telephone calls (either conventional or videoconferencing) at both offices on the overlapping days (Monday-Wednesday) will improve communication and avoid misunderstandings and misinterpretations. Second, creating a problem-reporting database that is accessible to both the client and the vendor's technical support staff will provide a common ground for problem reporting, discussion, tracking, and resolution.

While this example is an extreme case, it illustrates the importance of having trained, experienced staff available locally. If local support cannot be provided, the vendor should make support center staff available on a 24×7 basis if the product is deployed internationally. As much care should be given to evaluating support personnel and services as in selecting the software. If possible, talk to users of the same software at another company in the same area to gain their perspective.

User manuals/programming support. Documentation is one of the key factors in selecting an application, yet it is the last thing considered—if at all. Most users find that user manuals are woefully lacking, difficult to use, poorly indexed, and/or written by technical writers who are not users or geoscientists. Poor application documentation leads to frustrated users, which in turn leads to lower productivity.

Evaluation of user documentation should follow a formal and comprehensive plan, in much the same way the application itself is evaluated. This plan should address at least the following items:

- Is the table of contents (TOC) organized in such a way that the user can identify the desired section and general area of the manual quickly and easily?
- Is there a detailed index that provides cross-referencing to similar topics or keywords?
- Are the user manuals and related documentation available online in a universal, platform-neutral format [e.g., Adobe® Acrobat (.pdf) format]? If so, are these manuals indexed with hypertext links that allow easy movement from one section of the document to another?
- Does the documentation include cross-reference indexing to related topics and sections of the manual?

Training. Software training is a complex issue to consider, especially if the purchaser has little or no knowledge of the product. Most software vendors provide some basic level of training. In many cases, however, this training does not extend beyond a fundamental user awareness briefing. At that point, the company or individual must seek additional training or coaching, which may bear a significant additional cost.

Cost considerations. The cost of a software application is usually the first (or only) factor considered prior to purchase or renewal. Non-geoscience administrators tend to look for the lowest cost solutions, while the technical end user looks for maximum flexibility and functionality. Rarely are these the same product.

Therefore, it is very important that geotechnical users thoroughly evaluate all possible solutions. This evaluation should compare functionality and features objectively and critically, resulting in selecting the best possible solution. Best, in this sense, includes initial cost, maintenance fees, support factors, training needs, and other nontechnical issues in addition to the actual features and functionality that form the basis of the application requirements. Creating a feature-by-feature product comparison matrix (similar to software and hardware reviews in popular computer magazines) helps identify the best buy and will be very useful in convincing financial managers of the best technical choice.

GENERAL DBMS TYPES

Many different types of database management systems are available today. Selecting the system that is best suited to meet the objectives of the data management project requires a thorough understanding of the advantages and disadvantages of each.

Relational database. The relational DBMS is the most common and most flexible type of system available today. Basically, data are stored in fields or columns within individual tables. One or more of the fields is considered a key or index field and is used to link multiple tables together. Data tables are linked together by key fields in multiple tables that contain the same information. These links can be one-to-one, many-to-one, or one-to-many (see *Key Terms and Concepts* chapter).

The primary advantage of a relational system is that common data are stored together in tables that can then be linked together in various ways so data from many different tables can be viewed. Without the relational system, all the data needed for a particular view of the database would have to be stored in the same table (i.e., a flat-file database).

Hierarchical database. This type of database is best represented by the tree or folder structure, common to most computer file storage systems. Starting at the top of the tree, the database splits into multiple branches until the actual data or data files are reached. While an excellent method to store and archive data, this approach is not well suited for dynamic data retrieval and manipulation.

Flat-file database. The flat-file database is a collection of individual data files organized using some method of indexing. A flat-file database can be managed in some cases using a hierarchical method. Each flat file contains all information needed for a particular purpose or data type. Examples of flat files include LAS-format log data files, ASCII-format data files, and SEG-Y or other formats for seismic data files (SEG-Y is one of several data storage standards developed by the Society of Exploration Geophysicists). The largest disadvantage to flat-file data storage is the inability to easily update and modify the file contents. However, for certain applications (like storing LAS-format archival log data or SEG-Y data), they are ideal.

Proprietary database. Many existing legacy data management systems were developed using a proprietary data format, file format, and storage structure. Legacy data management systems present an unusual problem. On the one hand, they are typically extremely fast and efficient, and many users are reluctant to give up access to them. In most cases, however, modifications may be expensive or impossible, especially in complex systems where the developer is no longer available or the vendor is out of business. Furthermore, there may be no easy way to migrate the data from the existing system to a more modern, advanced system.

Spreadsheet-based database. The spreadsheet-based database is the most popular individual data management tool. Most PC users find the learning curve less steep, and the current generation of spreadsheets allows the individual to develop powerful data management tools that combine the best features of flat files and relational data managers. When combined with the virtually endless possibilities for customization and computation, the attraction is obvious.

While the lure of rapid development and on-the-fly customization draws many to this solution, too often companies and individuals try to stretch this development environment beyond the desktop level into a broader, more comprehensive environment. At that point, the spreadsheet database cannot compete with larger, more powerful systems that are needed to advance beyond the individual or small group level. For a small office or independent user, a spreadsheet-based database system may be the obvious and most cost-effective solution. Database design concepts presented here should apply nonetheless.

Object-oriented database. The object-oriented database was at one time the promise of the future. A complete discussion of the concept of object-oriented programming and object-oriented databases is beyond the scope of this book. However, the development of true, object-oriented data management solutions is only now beginning to have an impact on geotechnical database systems. The most successful object-oriented DBMS available for the petroleum industry is the GeoFrame®/Finder® product from Schlumberger GeoQuest. A complete discussion of the object-oriented database can be found on the POSC (Petroleum Open Software Corporation) Web site, as well as several examples of its application (see Appendix A). For the small or independent user, however, an object-oriented database may be impractical because of cost and complexity issues.

DATABASE APPLICATIONS

During the planning process, a great deal of consideration should be given to the database application that will manage the geotechnical data. An unwieldy or unsuitable database application will make it extremely difficult to provide the flexibility and data-specific features that geotechnical data require.

Unless a company has been operating in a complete technological vacuum, there will undoubtedly be some sort of DBMS already installed and used for other sectors of the business. Commonly, this will be for financial and/or engineering data management, but these systems may not be suitable for geotechnical data. While the temptation is great to use the existing system to take advantage of the economies of scale, several factors should be considered. These factors include flexibility (in terms of modifications and customization), the ability to link to geotechnical interpretive applications, and the ability to load and use a geotechnical-specific data model.

Flexibility in modification and customization

If an existing system is being considered for use with geotechnical data, it may require a certain amount of customization and modification so it can fully handle the specific types of data and user requirements planned. Also, if the system is primarily used for handling small, manageable, and relatively static volumes of data, it may not have the capacity to handle the potentially large volumes of geoscience data.

In most cases, it will be possible to determine an approximate value for the current storage capacity requirements and data processing speeds needed to handle existing data volumes and user demands. However, geoscience data (especially seismic data) are anything but static. While it may be difficult to predict the exact future requirements, input from users and corporate planning groups (if present in the company) can be helpful in making these estimates. Some of the factors to consider in making these estimates follow:

- Are there large development projects in progress or planned for the near future? If so, estimate the volume of new data from drilling, geology, wireline logging, and other sources.
- Are any currently held assets unexplored or underexplored? If so, what estimates can be made regarding the volume of 2-D and 3-D seismic that would be needed to explore these assets fully? Remember that even in mature areas, new 3-D or even 4-D seismic data may be added, and the data volumes of these surveys can be enormous.
- What is the potential for additional acreage (developed or undeveloped) in the near future? If the organization is relatively static, focusing mainly on development of existing assets, additional large volumes of data are minimal. However, if the company is involved in active asset acquisitions, the potential for data additions is considerable.
- Is there the potential to acquire other corporate databases because of a merger, acquisition, or joint venture? If so, not only do the new data need to be added to the existing database, but there will be other compatibility and data transfer issues as well.

An existing data management system, unless it is already being used for geotechnical data, is rarely directly usable for geoscience data. The ability to customize the system, either through built-in or add-on developer's kits or through additional optional modules, is a high priority. Without this flexibility, more time will be spent trying to make the system do something it was never intended to do than will be spent actually managing the data.

Ability to link to interpretive applications

Beyond the goal of loading data to the data management system, do not lose sight of the fact that the ultimate goal is to get the data into the hands of the users/interpreters. To do this effectively requires some level of data transfer or linkage between the database and the interpretive applications. If this capability is available in the existing system, an additional layer of programming effort is eliminated. Otherwise, the time, expense, and staffing requirements to create these links must be accounted for in the overall design of the database.

There are three primary methods of transferring data between a database and an interpretive application. The most effective and reliable means of moving data from the database to the application is through a *direct link between the two systems*. These features are available on the major software product suites available from Landmark and GeoQuest and as options on many smaller applications. Of course, this option usually comes with a price: you can sometimes only link between a particular application and the DBMS product provided by the same vendor. While more effort is being made to make these links cross-developer compatible, there is still work to be done in this regard.

Several *third-party software developers* have recognized the need for providing the links between the data and the application and have created products that move data from a database to an application (or the reverse), regardless of the vendor in either case. One such company is Oilfield Systems, Inc. Since third parties are not primarily in the business of selling DBMS products or geotechnical interpretive applications, they can often provide more flexibility and options than the application providers.

The simplest, oldest, and in many cases most reliable method of moving data between the database and the application involves a two-step *file transfer process*. First, the data are exported from the database into a flat file; then the data are imported into the interpretive application. If the existing DBMS does not have some sort of data export capability to create a file format that can be imported by the interpretive application, then the DBMS is entirely unsuitable for managing geotechnical data. If these capabilities are available, their limitations should be clearly understood and considered in terms of how much time and programming effort will be required to solve them.

Selecting a data model

Most commercial DBMS applications can support many different data models. The DBMS is the container that stores the data, while the data model defines where, how, and in what format the data are stored in the database. The data model also defines the relational rules that link tables within the database and provides a way to validate the data to ensure quality and consistency. Several large geotechnical data models are available, and one may be more suited to the needs and objectives of the overall project.

Commercial data models. By far the easiest method to implement and use a data model is to purchase a commercial model from a software vendor. The entire database schema, including table structures, indexes, and relational rules, are already created and ready to load data. By purchasing a commercial data model, the vendor generally provides support and maintenance for the model as part of a licensing and support agreement. Several commercial data models are available that can be applied directly to the petroleum industry.

The data model developed and supported by the Petroleum Open Software Corporation (POSC) is considered an object-oriented data model. While POSC has been a strong proponent for this model, it has taken a long time to develop to maturity and gain widespread acceptance by software vendors. A complete discussion of the POSC model can be found on the POSC Web site (see Appendix A for more information).

One of the earliest data models developed for the petroleum industry was the Petroleum Public Data Model (PPDM). Like POSC, the organization that developed this data model is made up of industry participants, including software vendors and large petroleum companies. PPDM is a very robust, stable data model that is ideally suited to the unique data types encountered in the geoscience and petroleum fields. A complete discussion of this model can be found on the PPDM Web site (see Appendix A).

The IRIS21 data model was developed originally by Petroconsultants (now part of IHS Energy). This data model was created to handle basic petroleum industry data as well as the more text-oriented information provided by Petroconsultants. Most of this information was oriented toward industry activity, competitor activities, and land/lease data. As such, this data model is not as well suited to storing dynamic geotechnical data as either POSC or PPDM.

Other commercial data models are available but are not listed here. Virtually any data model can be modified and extended to include geotechnical and petroleum data. However, if those capabilities are not included in the original version, the model will probably not be entirely suitable for most geoscience applications. There is also the issue of compatibility with application software, which is essential for the database project to succeed.

Case history: DBMS selection and customization

This case history illustrates the potential problems that can occur during the DBMS selection phase and during any subsequent customization of the data management system. While this case is an extreme example, the potential for these types of problems remains.

A large independent company was the operator for multiple, large gas fields containing hundreds of wells and over 100 active reservoir intervals. A DBMS was needed to handle the huge amounts of engineering and geotechnical data. Because of the international location, INGRES® was selected over Oracle® due to various support and training issues. At that time, the INGRES® DBMS software did not provide user-friendly interfaces, so an extensive, custom-written, menu-driven interface was developed that included a broad set of functions and report-generating capabilities.

As the data management industry matured, object-oriented programming (OOP) and GUIs provided more user-friendly interfaces with the data. When the petroleum industry (finally) embraced data management technology, the *de facto* RDBMS of choice was Oracle®. However, in this case there had been such extensive customization and dependence on the proprietary solution that it became virtually impossible to migrate the data to any other system or platform.

This case history illustrates the importance of attempting to make the best choices at the project outset in terms of software and degrees of customization. These choices then need to be reviewed regularly to ensure the original scope and objectives of the project are being met by the current solution (and to eliminate other developing factors, such as major shifts to other products). Had changes been made earlier in the life cycle of this DBMS solution, a great deal of time and effort would have been saved.

Developing a proprietary data model

Despite the availability of commercial data models, there may be circumstances (either financial or in terms of scale) that make it impractical to purchase and maintain a commercial data model. In these cases it may be necessary to create a data model from the ground up. This is no trivial undertaking, and a great deal of thought and planning need to done before embarking in that direction.

The foremost consideration in developing a proprietary model is the issue of maintenance and support. While it may be possible to have an individual or group plan and to develop and support a proprietary model, long-term support and maintenance become major issues should that individual or group be eliminated. Supporting and maintaining a complex data model require a thorough understanding of the model. If the original developer does not provide complete, detailed documentation of the model (which is commonly the case), it may be very difficult—or impossible—for someone else to take over the job.

Guidelines for proprietary data models. If a proprietary model is going to be developed, several points should be addressed early in the planning process:

Key fields should be carefully planned so all critical relationships between tables can be made without duplication of data in any tables (a fully normalized database).

Foreign keys must be included in all appropriate tables so links and relationships to the primary keys in other tables can be made.

All *relations* must be defined. After the primary and foreign keys have been identified, the basic schema for the database can be laid out. Before adding the parameter and data fields to the tables, it is important to make sure all key relationships are available and correct.

Once the keys and relationships have been created and tested on paper, *parameter and data fields* can be added. At this point, several new relationships will be identified, possibly requiring additional new key fields or entirely new tables. The whole process is then repeated.

When developing a proprietary data model, the designers should account for all of the different *types of geotechnical data* that will be encountered. While most data management applications routinely handle normal data types (text, numeric, logical, dates), the proprietary model should allow for special types that are common in the geotechnical world. These include large binary objects, image files, seismic data, and spatial data. A complete discussion of geotechnical data types is included in the *Geotechnical Data* chapter.

A proprietary data model must also be able to account for and handle the special problems associated with large *image files and spatial data files*. Image files include satellite imagery, scanned maps, and cross-sections. Large binary files include compressed-format log data files or any other large file in a binary format.

A *text-intensive database* presents several unique problems of data storage formats, query strategies, and reporting/exporting functionality. Large text fields, especially variable-length text data, require the use of variable-length memo or equivalent field types and present unique problems during data export, loading, and editing. It is also very difficult to write queries to search the contents of long text fields because the target text (such as keywords) may appear at any point in the text field. These queries are also very slow when searching large databases.

SELECTING A COMPUTING PLATFORM

In many if not most cases, the computing platform, operating system, and network protocols are already well established before a geotechnical DBMS is implemented. Unless the currently available hardware, platform, and operating system are totally unsuitable for use with geoscience database systems, it is probably best not to introduce major changes or upgrades at this point. The current hardware and operating system environment (as well as the current interpretive applications in use) have a direct influence on the platform used.

Application-driven databases

Geotechnical databases are often based on a suite of interpretive application software programs. In fact, most of the major petroleum industry application and interpretation software vendors (e.g., GeoQuest, Landmark) include data management as an integral part of their software. For some companies, especially those with limited resources, this type of database may be the perfect data management solution. Obviously, in these cases the computing platform and operating environment will be determined by the needs of the application suite and not the database.

The *individual or workgroup environment*, where small collective sets of geotechnical data are shared by several different disciplines, is ideal for focused workgroups. These groups can be multidisciplinary asset teams or individuals. The computing platform in this case is a single or at most a few UNIX-based workstations, generally connected by a LAN.

In most instances, the application-driven or project database runs on a *single platform*. The problems associated with sharing data across several different operating systems and running on different hardware are usually not found in this situation.

Independent databases also cause problems. The primary disadvantage of application-driven databases is that once the data are loaded to the project database, they becomes static relative to the corporate-level archival database (assuming there is one). Data are added, edited, and interpreted on the project database, while other editing and additions may be made to the archive or master database. Even if the edited and enhanced data are loaded from the project database at the conclusion of a project, it becomes difficult to determine which data are valid: the project data or the archive data. Most of the negative aspects of project databases result from attempting to upscale a database from a project level to a corporate level.

Application-independent databases

The application-independent database operates completely independent of the application software (although it may feed data to the application through direct links or through data file transfers). As such, this type of database can be installed on the hardware platform that is most suitable for the database without concern for the interpretive applications. However, data transfer issues are of primary importance in these cases.

The application-independent database is ideally suited for archival storage of data, especially for large corporate systems where the data are shared by a large number of users. In the past, large mainframe platforms were ideally suited for this purpose because maintenance, support, and control were centrally located. As desktop and networking technology improved, however, the high cost of maintaining and supporting mainframe databases was no longer economical in most cases.

One major drawback to a centralized database platform is the requirement for some means of moving the data to and from application software, normally located on desktop or workstation platforms. In many cases, these links must be custom developed, maintained, and supported. Unless the resources are available to commit to this work on a long-term basis, the use of a centralized database platform (such as a mainframe system) is not advisable.

IMPORTANCE OF STANDARDIZATION

Regardless of the choice of data model, computing platform, or database management software, the importance of standardization cannot be overemphasized. Adopting industry-accepted standards wherever possible allows easier upgrading, upscaling, and migration when and if it becomes necessary.

Upgrade and scalability issues

Choosing or creating a database system (consisting of platform, data model, and DBMS) that conforms to industry-recognized standards makes the process of upgrading or upscaling less painful. (It is still painful, but less so with standards in place.)

Upgrading issues. Virtually every software application, computing platform, and operating system will evolve through periodic upgrades and revisions. These upgrades include new functionality and features as well as (hopefully) corrections to problems with the current or earlier versions. Creating a DBMS that includes extensive customizations and modifications can make this process difficult or, in the worst case, impossible.

Modifications to the data model should be limited to essential extensions to the existing model. The creation of new tables should be limited as much as possible, and no modifications to the existing schema in terms of key relational fields should be made. All modifications should be thoroughly documented so they can be moved or migrated to any future upgrades to the data model. This is the only way the contents of the database can be migrated accurately and completely when these changes occur.

The future development and support of any operating system or platform database system that is dependent on a particular hardware platform or operating system (or that uses specific input, output, or storage devices) should be evaluated carefully. There are countless past examples of excellent data management systems and applications that became useless when the hardware or operating system they depended on became obsolete.

Scalability issues. Unless the organization's business model is to remain static or decline in size, data management needs will increase as the business grows. Even without significant growth, the ever-increasing amounts of available data will require upscaling the system at some point in the future. Several considerations in this regard are important:

- Can the existing or planned hardware platform easily and efficiently accommodate additional storage devices? As the volume of data increases, this will become a serious issue if the system cannot handle the additional storage resources.
- Will the DBMS accommodate the additional data volumes without serious degradation in performance? If not, will the system run on a multiprocessor platform or in a clustered architecture?
- Does the current network environment allow for additional users? How will the existing platform (including the network) adapt if remote users or field offices are added?

Accuracy issues

Developing, applying, and enforcing standards is the only way to provide data quality assurance to the users. If everyone is developing his or her own database, using different data models on different platforms, there can be no single set of definitive data that represents the best, most accurate, and most current information possible.

In a database system, there can be only two types of data: definitive and project. The definitive data represent the archival set of information that has passed through all the required quality control and data validation filters. In most cases, this definitive dataset has been approved by the user community as well. Once the definitive data are copied to a user's application, they automatically become project data. Project data exist completely outside the DBMS and are completely under the control of the user. As such, the quality and accuracy of that information are suspect until it has passed again through the same quality control and validation checks as it is eventually moved back into the database.

Data transfer issues

One of the most important parts of a DBMS is the smooth flow of information while transferring data from one point in the system to another. Data transfer is discussed in more detail in the *Data Loading and Input* chapter, but during the planning process, the basic standards for data transfer must be established. In most cases, the data being transferred should always come from the definitive data, not from project data.

Through the creation of single set of definitive, quality assured data, every user is guaranteed to get exactly the same information at any given point in time. The data should be delivered to the user in a predefined set of standard formats to eliminate the need for *ad hoc* export formats from the database. Custom-designed input templates for the user's applications are then used to load the data.

ROLE OF THE DATABASE MANAGER (DBM)

Only recently has the importance of the DBM's role started to gain the recognition it deserves. Previously, this role was viewed as a function or subset of the IT group and was staffed with programmers, database administrators (DBAs), and various other computer experts with little or no understanding of the underlying geotechnical data. With the recognition of the need for a database manager with formal geotechnical experience comes the problem of not having qualified individuals to fill these positions.

Managers and administrators

In the design phase of the database project, two key personnel roles need to be filled. The first of these positions is the more traditional DBA, and the second is the DBM. In fact, the DBM may already have been selected and currently leading the team designing the database project.

The DBA role traditionally has been filled by a computer expert with formal training in the design, operation, maintenance, and support of large-scale DBMS applications. This is a critical function, but DBAs rarely have the experience and expertise in the geosciences that enable them to see beyond the nuts-and-bolts aspect of the database and recognize the big picture that includes data, application, and user as a coordinated, related cycle.

A DBM is truly a unique individual. Some farsighted companies have selected these individuals from the ranks of the geoscience staff in exploration or development. In most cases, these individuals are the workstation power users or gurus who know a great deal about applications, data, and hardware in addition to their skills and experience in the geoscience arena. Qualification for this position should include as many of the following points as possible:

- Strong background in applied geosciences as a working geologist or geophysicist. Working knowledge of all phases of exploration and production, including hands-on experience with various workstation software applications.

- Expert-level proficiency in computer applications used in geoscience interpretation. Some direct experience loading data to the applications.
- Strong data management skills, including a working knowledge of large, enterprise-scale applications as well as desktop database applications. Some experience in writing and using SQL scripts, and familiarity with other major scripting and programming languages. Some experience with browser-enabled technology is helpful.
- Proven leadership skills and the ability to work easily and effectively with highly trained professionals from various disciplines. Excellent communication skills, both oral and written. Ability to communicate and present ideas effectively. The often underestimated importance of effective geotechnical data management requires a great deal of salesmanship.

General roles and responsibilities

The responsibilities of the DBM and DBA are determined to a certain extent by the size of the database, the complexity of the data, and the user community. This, of course, is directly related to the overall size and activity of the organization. However, certain general roles and responsibilities apply to any organizational scale.

DBM roles and responsibilities. The DBM's role and responsibilities in the organization can be subdivided into primary and secondary functions. The primary function of the data manager can be summarized as follows:

- Conduct or direct regular screening and evaluation of the data content, flagging any potential problems or errors for correction or follow-up with the appropriate users. Evaluate current data input validation workflow procedures and modify them as needed. If possible, work with available developers to create an expert system approach to data validation to automate the process as much as possible.
- Evaluate workflow processes with the users to better understand the critical flow of data to and from the applications. Develop and implement policies, procedures, guidelines, and standards to make the process efficient and to provide the highest quality data possible.
- Provide formal and informal training to users in terms of various data management topics, both one-on-one and during periodic training sessions.

Secondary responsibilities of the DBM are determined by the individual's level of expertise and maturity of the database system. Some of these responsibilities include the following.

- Work with the users and developers to identify inefficiencies, missing functionality, and areas for improved data validation. Thoroughly test and certify modifications and improvements to the interface before deploying the interface to the users in a production environment.
- Help the user coordinate data entry services, perform limited data format conversions, create project databases, develop workflow procedures, and perform specific data transfer and loading services, as required.

DBA roles and responsibilities. In contrast to the DBM, the role of the DBA is generally more a function of the IT sector of the company and is somewhat beyond the scope of this book. The general responsibilities of the DBA include the following functions:

- Maintain the overall health of the DBMS and its contents. This requires storage space optimization, performance monitoring and optimization, rebuilding tables and relational links where needed, and all the other activities required to keep the system up and running at all times.
- Maintain the overall security of the database. This includes granting (or revoking) user access to the database depending on authorization. Ensure the database is backed up routinely and regularly, including regular full and incremental backups so the database (or even specific rows in specific tables) can be restored. Schedule regular tests of the integrity of the backup procedures, including actually restoring selected data from backup to ensure the procedures function as planned and provide the desired level of protection.

Integration and coordination functions

The most important role the DBM fills is integrator and coordinator. This position is unique, in that it fills a large gap between the worlds of computers/IT professionals and the practicing geoscientist. This is not to imply that the working geoscientist is not technology oriented, but that a coordination (and communication) gap exists between these two groups. The IT professional generally is unfamiliar with the actual use of geotechnical data, while the geoscientist is not always familiar with the latest technology and nuances of the complex world of data management. Therefore, the DBM must be familiar with both worlds, preferably from having worked in both these functional areas. This makes the DBM a pivotal individual in the overall organization.

User communications functions

One of the most effective methods of communication today is through the Internet or company intranets. An excellent way for a project leader to provide immediate information about the status of the DBMS is via a Web site established for that purpose. While the actual design and implementation of the Web site depend on company standards or individual preferences, some basic content should be included on any site.

- General information on overall goals and objectives of the database project, including current development status, completion timing, and other key information
- All relevant documentation, especially with regard to the policies, standards, guidelines, and objectives
- Contact information (names, responsibilities) for all key individuals responsible for or involved with the project development
- A feedback mechanism to allow users and visitors to the site to provide comments, suggestions, and observations about any aspect of the project

During (and following) the database planning phase of the project, consideration must be given to the various data types and formats that will be handled by the data management system. The next chapter introduces the common geotechnical data types and formats, and some of the fundamental concepts of data storage and validation.

5

Data Types and Formats

One of the most difficult tasks in developing a new data management project or in migrating data from a legacy system is to clearly define and understand the wide variety of data types and formats involved. In many cases, especially with legacy systems, there is limited or unclear documentation regarding data formats and data types for the data stored in the system. This chapter reviews the most common data types and data formats, defines a general set of guidelines regarding data format standards, and introduces the concepts and methods of data validation. Examples of typical geotechnical data are used to illustrate these concepts and methods.

INTRODUCTION

Part of the planning process in database design must be devoted to considering the types of data that will be stored in the database. Although most geotechnical data are fairly straightforward and standardized through years of practical experience, individual needs vary. In part, data types and formats are determined by the user's data. Other considerations are dictated by the structure of the database management system (DBMS) application to be used. The selection of an appropriate DBMS application is covered in more detail in another chapter, but some of the factors to review are included here.

Scalability and portability considerations

Early in the planning process, careful thought should be given to the issues of scalability and portability.

Database scalability. For a small company developing a geotechnical database for the first time or larger companies considering changing to a new system, the issue of scalability is both difficult and important. It is easy to design a database system that will meet the immediate and known needs of the current users with the available hardware. However, as the system grows, hardware improvements are made, users become more sophisticated, and system usage increases (as well as data storage volumes). At that point, it becomes necessary to expand and/or enhance the system. The ability to increase the size, scope, and scale of a database is therefore critical. In addition to the choice of DBMS application and data model (see *Planning Database Projects* chapter), the following require consideration:

- Can the existing system accommodate additional data types?
- Can the system accommodate large increases in data volume or use without degrading system performance?
- Are there physical or practical limits to the basic structure of the database that would prevent this scalability?

Portability. No matter how carefully a system is planned and implemented, there may come a time when the database contents will need to be moved (migrated) to another system. By ensuring that the data storage formats use industry-accepted standards, this migration will be smooth and relatively painless. When proprietary data formats and proprietary data models are used, moving the data to a new system becomes difficult or economically unfeasible.

Modifications to commercial data type definitions

Although commercially available data models and DBMS applications are suitable for general purposes, geotechnical data have some unique aspects that may or may not fit the one-size-fits-all commercial database. While it is not always advisable to make significant modifications to a commercial product, there should be some capability in this regard or the functionality and usability will be compromised.

Data Validation and Exceptions

Data management projects tend to adhere to the 90/10 rule: It takes 10% of the time to get the data into the database and 90% of the time to ensure data quality and accuracy. Several general comments and recommendations can be made concerning data validation.

Use validation rules

The most effective method to provide quality control for data is at the point of entry into the database. Most DBMS applications that are appropriate for geotechnical data can provide several levels of data validation. (See also the discussion of data dictionaries in the next chapter.)

Table-based validation. The most fundamental method of performing data validation and screening is by using tools that are already embedded within the structure of the database tables. Most DBMS applications let the user specify the data type, minimum and maximum allowable values, data format, and other parameters for each element of the data table. Using these features, for example, prevents the user from entering text in a numeric field, numeric values that are outside preset limits, or any data that require a specific input format.

Form-based validation. An even higher level of control is provided when there is a custom input form that serves as an interface between the user and the database. A full discussion of this subject is provided in the *Designing the User Interface* chapter. Conceptually, the input form can provide predefined selection lists, templates for input formats, and even radio button or checkbox controls. These features of form-based interfaces not only speed the data entry/editing process but also provide a high degree of confidence that entry errors are eliminated before they enter the database. Form-based data validation requires an extra layer of programming to provide a human interface, but the tools to accomplish this are generally simple to use and apply. In many cases, these forms can be developed and maintained by nonprogrammers.

User validation control. Of course, the most effective method of data quality control is using experienced data entry personnel. There is no substitute for the data entry person having at least a working knowledge of the underlying data. Obviously, highly experienced geoscience professionals cannot be used as data entry clerks (although they are often the only people available to do so) because this takes them away from interpretive work. However, hiring individuals with a geotechnical background and providing appropriate, focused training and mentoring can eliminate data entry problems before they reach either the form-based validation or the data table validation layers.

Even with several layers of data validation rules, there is no way to prevent the entry of incorrect data (as opposed to invalid data). Using validation rules simply prevents entering the wrong data type, data beyond expected limits, or data in the wrong format. Other methods must be used to find and correct erroneous data.

Duplication or redundant data

Conventional database design rules dictate that data present in one relational table should never be copied or duplicated in another table. The justification for this rule should be obvious. First, this is a waste of database space; second, there always exists the possibility that one of the copied data elements will be modified and the other copy will remain unchanged. At that point, it is impossible to tell which version is correct.

In the special case of spreadsheet or flat-file databases, however, it is sometimes necessary to break the rules of database design and copy redundant data into a single table. This is necessary to make the flat-file data tables self-contained, since this type of database typically does not have the relational links between tables that a relational database does. For example, a typical database might contain well location (header) data in one table and stratigraphic tops in another table. Retrieving a set of formation tops for a specific well or group of wells is a relatively simple task using a relational database. The equivalent flat-file database would need to have all necessary well information stored with each formation top record to make it possible to load the same information into an application.

Storing derived data

As a rule, data elements derived or computed from other data elements in the database should not be stored in the database. This complex issue is hotly debated between data management and end users. In the purest sense, derived data can be created on the fly as data are retrieved from the database (especially for simple arithmetic calculations). Derived data that require more complex mathematical algorithms or sophisticated processing can also be done on an *ad hoc* basis. There are situations, however, where storage of derived data may be justified, as shown in the following case history.

Case history: formation tops in deviated wells. A table containing formation tops for a number of highly deviated wellbores will have the necessary measured depth value for the pick, as this is the observed data. However, to compute the true vertical depth (TVD) or true vertical depth subsea (TVDSS) values will require that the well location, elevation data, and well deviation survey data be retrieved. Then the TVD and TVDSS points for each point in the directional survey need to be computed using a complex algorithm involving spherical trigonometry. Finally, the values of TVD and TVDSS need to be computed either through the same algorithm (as pseudo-deviation points) or through interpolation from the directional data.

A much simpler approach would be to perform that calculation only at the time the formation top is entered or edited and stored in the tops database. That way, no matter how many times that top is retrieved and used, no calculations are involved in the retrieval. This may improve performance significantly.

Many of the raw data elements contained in a geotechnical database are used in subsequent computations. As shown by this example, there are instances where storing a computed field is efficient and saves a great deal of programming effort. However, storing data in computed fields should be avoided unless necessary from an operational or programmatic standpoint.

COMMON GEOTECHNICAL DATA TYPES AND FORMATS

There is a wide variety of geotechnical data types, and multiple data formats found within each data type. The most common data types and formats are covered in this section. Specialized data types and formats that are unique to certain applications are beyond the scope and intent of this book.

Character-based (text) data

Despite the volumes of numeric data used in geotechnical applications, by far one of the most common data types is basic character data. Almost every data table and all databases have some character data. Unfortunately, character (text) data is also one of the most misapplied data types in geotechnical databases. To avoid the pitfalls associated with misapplying text data, consider these rules-of-thumb.

Don't use character data for numbers or dates. When storing date information in a database, the most common mistake of this type involves storing a text string that represents a date. That is, the actual date field contains a string of characters like 03/18/97. While that string of characters may accurately represent the date (March 18, 1997), a character string cannot be used in any computations, and the display format cannot be changed. On the other hand, specifying that a particular field is a date field provides complete control over the date information being stored there:

- Invalid dates cannot be entered.
- The display format can be changed easily.
- The date can be manipulated and used in computations.

In some databases, numeric data are represented as text strings. This is especially true of older, legacy database systems. While the value appears as a number, several problems will result from this approach. First, there is no simple way to perform a mathematical computation using text data as input. Although there are database functions that allow a text value to be converted to a number, this adds complexity to the computation. Second, numeric indexing cannot be done with text-type data (although the text can be sorted, it does not always yield the same result at a numeric index).

For example, if well names include a well number as part of the text string, sorting the names will not necessarily put the wells in sequential order. Consider Figure 5–1 where the well names and numbers are included in the same text string. The records are shown sorted by the field WELL_ID_1, in which the well identification (ID) code includes the well number as an unpadded numeric character. The resulting sort order produces a list of wells that are not in the desired numeric order (i.e., all the wells that have "1" as the first character are sorted first, regardless of actual numeric order).

TABLE: WELL_MASTER		
SORTING EXAMPLE: SORT BY WELL_ID_1		
WELL_TEXT	WELL_ID_1	WELL_ID_2
Well Number 1	WELLNAME_1	WELLNAME_001
Well Number 10	WELLNAME_10	WELLNAME_010
Well Number 11	WELLNAME_11	WELLNAME_011
Well Number 2	WELLNAME_2	WELLNAME_002
Well Number 20	WELLNAME_20	WELLNAME_020
Well Number 3	WELLNAME_3	WELLNAME_003
Well Number 4	WELLNAME_4	WELLNAME_004

Fig. 5–1 Well Name with Unpadded Numeric Text. Because there are no "leading" zeros between the well name and the well number, the sort order is not correct.

However, by "left-padding" the well number with "leading" zeros (so that well "1" becomes well "0001," the resulting well ID code (as shown in field WELL_ID_2 in Figure 5–2), sorting on that value will yield the intended and desired results (see Fig. 5–2).

TABLE: WELL_MASTER		
SORTING EXAMPLE: SORT BY WELL_ID_2		
WELL_TEXT	WELL_ID_1	WELL_ID_2
Well Number 1	WELLNAME_1	WELLNAME_001
Well Number 2	WELLNAME_2	WELLNAME_002
Well Number 3	WELLNAME_3	WELLNAME_003
Well Number 4	WELLNAME_4	WELLNAME_004
Well Number 10	WELLNAME_10	WELLNAME_010
Well Number 11	WELLNAME_11	WELLNAME_011
Well Number 20	WELLNAME_20	WELLNAME_020

Fig. 5–2 Sorting Example Using Left-Padded Numeric Text Strings. By adding the "leading" zeros between the well name and the well number, the sort order is numerically correct.

Textual representation of numbers does not allow easy reformatting of the data for display. For example, if three numeric values are stored as text strings as 123.456, 123.5, and 123.4567, there is no simple way to provide consistent display of the data with a common fixed decimal format or to provide rounding functions.

Consistent case and spelling. When text strings are entered into the database, it is extremely important that all the information be spelled correctly and entered in a consistent case format.

As an example, if the data for a particular well name are entered with inconsistent capitalization, later queries on the data may not locate all desired records. If the well name is entered as Alpha No. 1, ALPHA NO. 1, and alpha no. 1, a simple search of the database for information on Alpha No. 1 may only find one of the three records unless a more sophisticated search is performed.

At some point in school, children are told that spelling counts, and it is even more critical in technical data management. Simple spelling errors can result in valid data not being located by search tools. Consider a case where the search field is the well operator and the data have been entered as Exxon, Exon, Exxonn, and Exonn. Obviously, a search for one of these names will not find the other three.

Case history: specialized numeric formats. A vendor-provided commercial database provided latitude and longitude coordinates of well locations. The location coordinates were stored as text strings (see above discussion on this problem) in the format 109° 23' 14.623". The justification, as explained by the vendor, was that it was easier (for them) to generate reports and maps where this information could be displayed/printed without having to convert various numeric data into a combined text string. Of course, the data management staff and end users had to perform a great deal of work to convert the text data to numbers so they were usable as spatial location data. Other problems associated with special characters and punctuation arise when the characters are unprintable on another system or result in unwanted control commands being sent to the output device.

Query considerations. If a particular data field will be used as part of routine searches (queries) of the database, some consideration should be given to the format of the data. For example, if a table of formation names will be queried and reports will be generated that list formation names, the planning needs to include formation-naming conventions. For example, using names like

"Top of Unit A" and "Base of Unit A" will result in the two different records for that unit being displayed in different parts of the report (i.e., all names that start with "Top…" will be together and all names that start with "Base…" will be together). This may not be the desired result. By changing the nomenclature slightly to "Unit A Top" and "Unit A Base," the formation names will sort in the correct order, with the top and base of a unit adjacent in the sorted list (although in this example the base will be listed before the top!).

Form and/or table validation methods. If a field has the potential to be used as part of a user-defined query, free-form text should be avoided. Instead, use a predefined (or content-defined) picklist or drop-down list from which the user must select a text string.

Free-form text fields. This format is most often used for comments or remarks. Fixed-length text fields should be avoided unless required by the DBMS application. Regardless, never use multiple records to store variable-length text fields. Also consider using DDE or OLE links to files outside the DBMS for text documents longer than a normal paragraph. Most current DBMS applications have a memo data type. This is a variable-length data type specifically designed to handle this sort of information.

Case history: storing numbers as text. This case history involves one of the worst violations of the basic rules of data storage: storing text strings in a data field that would normally contain numeric data. In this instance, database users from different departments and disciplines required the same information (in this example, location coordinates) in different presentation formats. Although the original observed data were correctly stored in numeric data fields, various users required custom formatting, usually for reports and data listings.

For example, a well location might be at 12° 32' 26.789" (stored in the database as decimal degrees, 12.54055°). Because the original location was surveyed in degrees, minutes, and seconds (DMS) format, both the DMS format and the decimal degree (DD) format versions must be stored. In addition, the custom formatted version (with appropriate degree, minute, and second superscripts) was needed for several report formats. As a result, there were as many as 15-20 different fields in the database. Whenever a location was edited, each field had to be checked manually.

In this situation, a review of the input data accuracy was conducted to ensure that any conversion from DMS to DD format would preserve the original data accuracy. Next, the storage format was standardized as DD so the location information could be used directly in computational functions without requiring conversions. Finally, the custom report formats required by each department were designed into the report or export part of the DBMS. In this way, the custom formats were created on the fly rather than stored in the database. The result was a more elegant solution that required lower database overhead and eliminated problems associated with synchronizing the multiple fields and formats.

Numerical data

Numerical data are the most common data type in most geotechnical databases. Because of the wide variety of numerical data and the wide variation in applying the data, a great deal of consideration should be given to planning the portions of the database that include numerical fields. By far the most common mistake made with numerical data fields is the use (or misuse) of significant digits. Surprisingly, many people (nontechnical users, in most cases) honestly believe that by increasing the number of places behind the decimal point, the accuracy or precision of the data is increased. Conversely, data modelers without a geotechnical background sometimes inadvertently round off data by reducing the number of significant digits. This can be extremely dangerous when dealing with certain data types.

Surface location data can be expressed in many different formats and projections. If the coordinates are stored as a Universal Transverse Mercator (UTM) value (in meters), retaining more than two significant digits (i.e., accuracy to the nearest whole millimeter) is unnecessary. On the other hand, if the coordinates are stored as decimal degrees, eight significant digits may be needed to retain the full precision of the original data. This is particularly true when conversions are made from one coordinate system to another to ensure the reverse conversion does not suffer from an unexpected rounding error.

Computational vs. noncomputational fields. Numerical data are often stored as text, even in spreadsheet databases. If the data are stored this way, it is very difficult to use them as part of a computation (sums, averages, etc.), and they can sometimes be misinterpreted in a query. If a numerical value needs to be used as a number, store it as a numeric field (and not combined in a text field).

Numeric units fields. Regardless of the data type and format, it is very important to include information about the units in which the numeric data are measured. It is not sufficient to assume that all well-related data are measured in feet or meters (depending on location); there must be information stored in the database that specifically defines the units of measurement. If the numeric data are ambiguous, a units or dimensions field should be included on the same record. Dimensionless values, of course, are the exception.

Date and time data

While not strictly geotechnical in nature, date and time data play a critical role in most data management systems. Spud dates, completion dates, production intervals, circulation time—all must be captured and manipulated in some way. From a database perspective, dates and times that a record, table, or individual data element adds, modifies, or deletes from the database are very important. Thus, standardizing the format for date and time data becomes a very important and often overlooked part of the data management process.

The most common problem when dealing with time/date information is when this information has been stored in a database as text (character based). Consider the example table of date information in Figure 5–3. In this example, two different problems appear. First, there is no standardization of format (discussed in more detail later in this chapter). Second, the dates must be reformatted to make them useful or usable in a computational manner. Remember that the purpose of maintaining date information in many cases is to perform some sort of mathematical comparison or other computation.

Spud Date
January 1, 2000
1-1-2000
2000/1/1
01-Jan-00
Jan 1, 1900
1/1/00
11-October-2000
11/10/00
10-11-2000

Fig. 5–3 Example of Inconsistent Date Format. Using inconsistent data formats can create confusion, ambiguity, and lead to incorrect computations on the date values.

Recommended procedures for date and time data. The best way to deal with date information is to capture the data in a standard format that provides an unambiguous and easily interpretable format. Furthermore, a standard format allows much easier conversion and manipulation of the data.

Solar calendar or Gregorian dates are typically used as a standard internal storage format for many databases and spreadsheets. The calendar used as a base reference is not commonly considered when dealing with dates. However, special circumstances may require that calendars other than the widely used sidereal or solar calendar be applied. This would include any instance where Islamic Hijira dates must be stored or interpreted. The Islamic calendar uses a different starting reference, and Hijira months and years are shorter because they are based on lunar, rather than solar, periodicity.

If date information is not stored in a database that allows internal date/time storage, dates should be stored as text fields using an ISO 9000 standard format:

- Date sequence stored as Year, Month, and Day. This eliminates the problems associated with the many different date formats used in Europe, Asia, and North America.
- Year values, using four digits ("padded" with leading zeroes if required).
- A hyphen used to separate Year, Month, and Day.
- Day and Month values less than 10 should be "padded" with leading zeroes.
- Dates before Year 1 stored with a leading hyphen to denote a negative date.

With this discussion in mind, let's revisit the earlier example. As Figure 5–4 shows, the various inconsistent dates are now shown as a Standard Date that complies with the general ISO standard format. This illustration also shows the text-based date information in the Spud Date field copied into a new field, defined in Date/Time format. Note that the various dates have been reformatted into the internal format of the DBMS, which is not necessarily compatible with the ISO standard. The advantage of the internal storage format, however, is that the dates are stored as numeric values and can be used in computations without further reformatting. Choosing the right standard is not necessarily the same as choosing the most appropriate standard.

Spud Date	Standard Date	Internal Date
January 1, 2000	2000-01-01	1/1/2000
1-1-2000	2000-01-01	1/1/2000
2000/1/1	2000-01-01	1/1/2000
01-Jan-00	2000-01-01	1/1/2000
Jan 1, 1900	1900-01-01	1/1/1900
1/1/00	1900-01-01	1/1/2000
11-October-2000	2000-10-11	10/11/2000
11/10/00	2000-11-10	11/10/2000
10-11-2000	2000-11-10	10/11/2000

Fig. 5–4 Example of Consistent Date Formats. By storing dates in a consistent and predictable format, there is no ambiguity or confusion, and mathematical computations using the dates can be done more easily and more accurately.

Logical data

Logical, or Boolean, data types are useful but often misunderstood. In many database applications, it is helpful to have true/false or yes/no information stored to simplify and speed queries and reporting functions. However, in many cases the database developer or user simply adds a text field and stores the text strings as Yes, No, etc. At a query level, this is *not* the same as using a Boolean data type. Logical data (T/F or Y/N) is most easily utilized as part of a Boolean operation. To make most effective use (and to ensure proper portability and scalability), a Boolean or logical field type should be used. Exercise caution when dealing with logical fields in Visual BASIC, however: the digital value for true is –1 and false is 0 (zero).

Binary data

The use of binary data types has become more widespread and necessary as geotechnical information (and hardware) has become more sophisticated. The binary data type allows a database to store a much wider variety of information, making it possible to store and manage virtually any type of information.

Image data are a common example of a binary data type. They include virtually any information stored in a binary image format. These image files can be generated directly from an application (screen images, digital satellite imagery) or through scanning (maps, cross-sections, photographs, logs). There is a wide variety of image formats, and a complete discussion of the merits and use of these formats is beyond the scope of this book.

A binary data type can also contain the binary contents of virtually any file that can be stored on a computer system (with a few exceptions). Thus, it is possible to store grid files, special format files, system information, executable program code, and any other information of this type. By placing it in a binary field in a data table, the original contents of the file remain unchanged, but the information about the file (stored in either text or numeric fields in the same record as metadata) can be managed easily. When it is impossible or impractical to store the actual file in a binary field, it is sometimes possible to store a "link" to the file (the physical or virtual location of the file on the computer system) in a text field within the database (see Internet address links, below). Using this approach can lead to problems; if the actual (physical or virtual) disk locations of any of the files in the system are changed, the file "links" stored in the database are no longer valid.

Some compressed binary formats cannot be stored in a binary database field or they may become corrupted when doing so. It is always best to test the integrity of a unknown binary format before and after retrieval from a binary field to ensure that the file has not been corrupted.

Internet links are a special case of a file link that allows a direct link to an Internet Web site specified in the hypertext transfer protocol (HTTP) link address (e.g. http://www....). While very useful, the life expectancy of this type of information is rather short because of the dynamic and flexible nature of the Internet. If this data type is used, periodic checks of the validity of the HTTP links should be performed. This can be done using automated tools, but these tools may not be standard options or functions in most DBMS applications.

While it is possible to store virtually any information in a binary field, there are limitations and problems associated with the storage of executable program files. Most sophisticated software applications are not simply a single executable file but contain numerous related files, dynamic link libraries (DLL), configuration files, and other files. These related (and necessary!)

files are commonly stored in multiple, separate locations in the computer. The user must exercise extreme caution when storing executable files to ensure that all related files are included, either in a compressed archive file or as a self-contained installation file. Obviously, the same caution must be used when dealing with any special format files where other information or related files are needed to make the file usable.

With the foregoing introduction to common data types, formats, and potential problems in mind, the next chapter will focus on the actual design of the database system. Specific geotechnical data types and typical data formats are discussed in more detail in a later chapter.

6
Designing the Database

Even with established, functioning databases, there are many places where good database design (or modification) can result in better, more accurate, and more effective data management. While it is always best to design the database in the planning and development stages, any system can be modified and improved. Therefore, even when dealing with a commercial database product, many of the concepts introduced here will be important in designing, populating, modifying, and using a DBMS. This chapter is devoted to the actual design of a technical database, including planning for the types of data to be stored, how to deal with user activity and record deletions, and the use of CASE (Computer-Aided Software Engineering) tools to design and develop the project.

DATA DICTIONARIES

A data dictionary is a roadmap of the underlying database. In it are stored the names of the tables used, the columns (elements) contained in each table, and the data types and formats for each element in the database. In addition, the data dictionary shows the fundamental relationships between tables, primary and foreign index keys, view definitions, validation rules, and data relation diagrams. The data dictionary is the fundamental reference for the database and forms the basis for implementing and maintaining the database.

Importance of data dictionaries

The use of data dictionaries has traditionally been limited to the programmers and DBAs responsible for maintaining the database and its contents. However, having a fundamental understanding of the structure and purpose of data dictionaries can help the working geotechnical professional gain a better understanding of the database.

The advantages of using a relational database lie with the ability to provide links or relationships between the tables. Linking a key field in one table with a key field in another table creates these relationships. A data dictionary identifies which fields in a table are primary keys, which are used as foreign keys, and how the relationships are defined. This information helps the data modeler and DBA (and in some cases the user) map the relations between tables and find possible inconsistencies or places where the database can be optimized.

Certain fields in a table (especially those used as primary or foreign keys) must contain valid data before new records can be added or existing data can be edited. If a required field is not present, relations between tables cannot be made and processing errors will result. Generally, a field that requires some sort of valid data entry is shown as *not null*, meaning it cannot be left blank or null (empty).

One key use of data dictionaries is to provide internal consistency throughout all tables within the database. This ensures that all data of a particular type are contained in fields that have the same internal format. For example, if formation name information is contained in a 15-character field in one table but is related (through a foreign key relation) to the same information in another table where the field size is only 10 characters, a potential problem in the relationship can arise because the additional 5 characters will be truncated.

As part of the internal validation of data, it is important to set limits and ranges for data to be stored in each field. For numeric data, upper and lower limits (as well as default values) can be provided to reduce the possibility of invalid data being loaded to the database. Validation ranges cannot prevent errors, but they will prohibit any invalid data that do not meet the preset limits. For text or data information, requiring that the user input the data using a predefined input format or template can also be used as a first line of defense against invalid data entry. For example, if the well name format is specified as a combination of field name, well number, and sidetrack number, an input format could be specified as XXXXX_nnn_nn so that only 5 text characters (for the field name), 3 digits for the well number, and 2 digits for the sidetrack can be entered.

ORIGINAL VS. DERIVED DATA

While some users debate the actual definition of "original" and "derived" data, for the purposes of this book, original data will refer to the actual observed, recorded data prior to any modification, editing, or interpretation. Derived data, therefore, is any data that have been edited or modified in any way.

What data to store

There is a common misconception among many geoscientists that all data should be stored somewhere in the database. Considering the huge volumes of data associated with most projects, especially large field development and simulation, the loading, editing, and storage of any nonessential data become impractical and cost prohibitive. Even as unit storage costs come down with improved technology, storing nonessential data degrades application performance and data access times. The additional time and personnel required to load the data also become an issue. Therefore, it is very important to recognize and store only the essential data without losing any key data elements.

Fundamentally, the only data that should be stored in a database are the original, observed, or measured data as recorded or collected at the source. While this is a rather simplistic rule, it can be applied in most cases. There are, of course, exceptions to every rule. If the derived data require a very complex mathematical solution to generate, it is sometimes better in terms of system performance to store the derived data rather than require that the system perform complex solutions during data retrievals.

What data not to store

As a rule, data should not be stored in a database that can be easily derived from the individual elements within the database. In most cases, inefficient and denormalized data tables are littered with information that should be relegated to summary reports or simple data retrievals. To illustrate the effects of storing derived data in a table, consider the following examples and illustrations.

Net pay/net sand summaries. Geoscientists always summarize numeric data, especially when it comes to gross and net thickness of hydrocarbon-bearing intervals. No matter how detailed the investigation, the first question is always "How much?" or "How many?" It is natural, therefore, to want to have the subtotal and total thickness of these various parameters in the database.

There are two major problems with storing summary data in the database. First and most important, the total values are always based on summaries derived from multiple records elsewhere in the database. If the summary or total is stored in a record and the underlying data are subsequently changed, the summary data are no longer valid. Second, it is not always intuitive which records in the database have been used to produce the summary data.

Subtotal and *grand total* data are effectively the same problem as any summary data. Because changes in the records that are used to create the totals or roll-ups can change the total, these data should not be stored. Instead, a query or report should be created that generates the summary or totals on demand. In this way, the summary report always reflects the current contents of the records being summarized.

Another problem associated with summary values involves data that can be easily computed from other data in the same table. In many cases, simple computed values are generated from data in a table. The temptation is to store this information in another field in the same table. A good example of this problem can be illustrated with computed petrophysical data. If a table is used to store the results of basic log calculation information such as porosity (φ) and water saturation (Sw), often the log analyst will also store the hydrocarbon saturation ($1 - Sw$) and the bulk volume values [$\varphi (1 - Sw)$]. These derived values are again dependent on the values in the porosity and water saturation fields; so any time those values are edited, a corresponding update of the derived fields is required. As is the case with most derived data, it is generally better to create the derived data using a stored query or reporting function.

History Files and Deleted Records Files

History files and deleted records files are underutilized or often ignored in a database project. A history file is an activity audit that tracks what information was changed, added, or deleted from the database, by whom, and when. This provides valuable information about the activity in the system and can be very useful in recovering from potentially disastrous mistakes.

History files and deleted records files serve a twofold purpose. First, a history file maintains all information on user activity on the system. Second, using history files and deleted records files, the database can be "rolled back" to a condition before the changes were made (similar to the "Undo" feature or button on most spreadsheet and word processing applications). Any time a user edits, adds, or deletes a record, the system effectively logs the changes into a history file. Some systems refer to this as a log file. When data have been changed, it is important to know who made the changes, when the changes were made, and what the original values were. Ideally, a history file should also allow a user to record comments (also stored in the database) that indicate the reason for the changes. This provides a somewhat permanent record of the origin and evolution of the data.

A deleted records file is a repository for any records deleted from the system. The file serves the same function as the wastebasket for deleted files and folders on a PC or Mac. Until this data record wastebasket is emptied, any deleted information contained within it can be recovered and restored to its original location. Working in conjunction with the history file and activity log, data removed from the database can (in most instances) be restored accurately. These tables of deleted records should be maintained for a certain period, depending on the level of editing activity, the sensitivity of the data, and storage space requirements.

Case history: deleted records history file. While each DBMS has unique features, most applications have some sort of "rollback" option that restores a table to a previous condition or state. Normally, this feature is used as an undo operation when an editing or updating process fails or is terminated unexpectedly. This case history deals with a situation that is more insidious and has the potential for creating massive problems and requiring a significant amount of manual work to recover (if, in fact, recovery is possible).

In this example, an inexperienced geologist at the international office of a large independent company was working on stratigraphic correlations in a giant gas field. The proprietary DBMS allowed any user with appropriate authority permission to change data and assigned the user's ID and a change date to the modified record. This individual, however, spent several weeks recorrelating the formation tops on hundreds of wells. During that time, new data were added, existing data were deleted, and dozens of changes to information other than formation tops were made. Unfortunately, when he completed his work, it was discovered that he had made massive errors in his fundamental assumptions and all the new tops were invalid.

At this point, the IT support staff recommended restoring the database tables that had been changed during the last three weeks using a regular backup set (another situation where you can't have too many backups). Since other users had been making changes to the same tables during this time, however, this solution was not considered practical. The geologist in question had done most of his work without written notes, so it was impossible to recover manually from the mistakes. Fortunately, the DBMS had a feature that copied the original version of a record to a history table before committing editing changes to the working table. With a minimal amount of SQL coding, the bad data that had been added to the table were removed and the good data were restored from the history table using a combination of edit date and user ID. In this way, changes made to other records during this same time (or even the same records by other users) were unaffected.

In an ideal world, a DBMS would have an automatic rollback feature to recover from complex mistakes such as this. However, accounting for this sort of problem and building in a deleted or edited history table provides a much easier path for error recovery. Although database backups are an essential part of any system, the backup datasets are mainly for disaster recovery—not for correcting careless mistakes.

CASE Tools

Computer-Aided Software Engineering (CASE) applications can be extremely beneficial in the design, maintenance, and operation of large, integrated relational databases. Although these techniques are more suited to the realm of extremely large applications, the concepts are presented here as an overview.

Definition and application

CASE tools are very powerful and popular methods of database design used by professional database modelers and designers. Imagine attempting to build even a small shed without adequate blueprints and specifications. Constructing a complex database is no less difficult, yet many developers attempt to do just this without the proper tools.

CASE tools are, in the simplest sense, a means of mapping the complex interrelationships within a relational database system. Beyond the simple construction of a series of graphical representations of the relations and links between tables, it is also necessary to evaluate relations, links, and indexes to ensure the database is "normalized." That is, the data table relations should not include circular references or other design problems that will ultimately create serious problems with the database as a whole.

Use during development

During the design phase of a database project, it is very important to develop and maintain standards for virtually every phase of the project. As the project evolves and becomes more complex, it becomes increasingly difficult to keep track of field names, data formats, primary and foreign key fields, and the complex relations between tables. CASE tools track all these items and have the additional benefit of providing design evaluation and analysis.

Unless you are dealing with a major data management effort and have the services of a professional data modeler, CASE tools may not be the ideal solution for you. In many cases the results can be accomplished through other, less expensive and less complicated methods.

OTHER DATABASE TOOLS

For small and intermediate database applications, there are other less rigorous methods of mapping a database. Most of these consist of database summary reporting methods or purely graphical methods.

Built-in database analysis. Many database applications have build-in analytical tools. MS-Access®, for example, has database analysis utilities that provide a complete printed summary of data tables, relationships and links, indexes, and other key summary data. While this utility does not provide the same level of analysis and control as CASE tools, it is not intended for the professional data modeler. For a quick summary of the database and a passable data dictionary, this is an alternative to more sophisticated (and more expensive) applications.

Graphical methods. In most cases, it is very beneficial to have a graphical representation of the data tables, with connectors showing the links and relationships between them. A number of low-cost solutions are available that will let the data manager plan and evaluate an overall database design.

Once a stand-alone product, Visio® is now incorporated as part of the MS-Office® family of integrated PC applications. Visio® provides the user with a customizable work surface and a wide variety of predefined templates or stencils that are used in much the same way that physical templates are used to draw symbols. The advantage, of course, is that objects can be moved, revised, and rearranged quickly and easily.

Although MS-Word® has a number of drawing tools available, the program functions primarily as a text processor. Many of the same functions and features found in Visio are available in a limited sense in Word, but this is not the ideal environment for building and editing graphics.

CUSTOMIZING COMMERCIAL PRODUCTS

Not all data management projects are developed from the ground up. More commonly, some sort of existing DBMS is already developed and installed, with or without a user interface. In these cases, additional steps must be considered.

Most commercial DBMS applications provide a more or less complete solution. In many cases, however, some degree of customization may be necessary to provide the user with an appropriate interface, to make changes to the underlying data model, or to provide custom loading, editing, export, and reporting functions that are not available in the off-the-shelf product. No matter what changes are contemplated to a commercial system, they should always be viewed as temporary solutions to an immediate problem. If changes are made, the vendor should be notified of the purpose of the change and provided with the details of the solution. In this way, future product releases can incorporate these custom-developed solutions when appropriate.

Case history: disposable code. Most custom software applications or utilities require a significant amount of expertise and time to create. It is therefore natural to have a certain pride of ownership and reluctance to abandon an application once it is created. As discussed in the text, there are numerous reasons to avoid custom software development. However, few off-the-shelf products provide a complete solution for every situation. In those unique cases, developing a disposable custom solution is a viable alternative.

An international operator needed a rapid method for loading, quality control, reformatting, and export of mission-critical data from a central database. The key data needed by the users included well locations, borehole trajectories, and formation top data. These data were needed in a variety of formats for numerous applications running on several platforms. The existing commercial RDBMS provided very few of the needed options and had limited customization capabilities. Solutions were needed within weeks, not years.

The solution to this case lay in the use of disposable or core code. Using the basic data model as a framework, a simple text-only, menu-driven interface was quickly developed. Next, the essential functions and computational routines needed to load, view, and edit the mission-critical data were written. Finally, export modules (essentially custom report generation formats) were developed. This process was completed in a matter of weeks, and data loading, QC, and export (for key data) capabilities were fully functional in 2-3 months.

As the basic operational system was deployed to technicians and the project moved from data loading and QC into a maintenance and improvement period, a more user-friendly GUI interface was developed. Using similar programming tools at this point allowed the developers to reuse sections of code that were developed in the initial version. After testing the GUI-based version, the older version was abandoned.

To be truly effective, custom-developed code should be used (where possible) to provide functionality not available in commercial products. Working directly with the vendor (in some cases through commercial alliances), the proprietary custom solution can speed the development of new features in the commercial product. Once developed and maintained by the commercial vendor, however, the original proprietary version should be abandoned.

Data model extensions

No matter how extensive a data model is used for the database system, certain specialized data types or specific operational needs may require the storage of data that was not anticipated by the data model designers. When this situation is encountered, it may become necessary to extend certain tables in the data model by adding new fields. It is never advisable to change field names, relational links, or data types for existing elements of the data model because this can have far-ranging effects that may not be obvious without extensive review of the data schema.

Extensions to an existing data model may be required when there is no predefined field in a table to store a particular data item. For example, if it is necessary to store special computed fields or comment fields that are unique or proprietary, data model extensions are needed.

The primary advantage gained by using data model extensions is that they provide an unlimited level of personal or corporate customization and modification to suit the particular demands of the data. The major drawback to custom extensions is that future upgrades or changes to the base data model (assuming it is provided and maintained by a vendor) will probably not incorporate these custom extensions. When changes or upgrades to the base data model are made, it is very important that custom extensions be added to the new data model before the data are migrated to the new data model version. Otherwise, the information contained in the custom fields will be lost. This again illustrates the importance of documentation and the use of a data dictionary.

Extensions to commercial data models typically include fields that would violate fundamental rules of good data modeling. For example, a particular table for directional survey data may contain only the normal measured data, without any computed data (see the detailed discussion on *Directional Survey Data* in the next chapter). However, the particular operational needs of the user or company may require that the computed data be stored with the observed data. This would require adding fields to the directional survey table to account for the computed data.

Generic Data Tables

Geotechnical data models are designed to store a wide variety of data types and formats and to account for most of the data encountered. However, there are exceptions to every rule, and a situation may arise where there is no defined data table available in the data model. In these situations, it is possible to create a generic data table that can be used to store virtually any data type or format and can be easily migrated to a normal database table if it is later added to the data model.

The generic data table uses a parameter-based approach, where each data value stored in the table has a parameter name, parameter value, and other optional values, depending on the complexity of the situation. The structure of a generic data table consists of the following fields (as a minimum):

- Parameter name field. Normally, this text field contains the name of the parameter that is being stored in the table.
- Parameter value field. This data field contains the actual data being stored in the table. The parameter value field must have an internal storage format consistent with the data being stored. However, multiple data types can be stored in the same table by defining parameter value fields for numeric, text, binary, or other data types to be stored in the same table.
- Parameter units field. This field is optional (though highly recommended), and used to store the units of measurement for the parameter. Obviously, if all of the data in the generic table are text format, this field can be omitted.

Generic data tables organize small amounts of data that otherwise would have to be hard-coded into an interface form (see chapter *Designing the User Interface*) or don't have a defined place for storage in the data model. To use the generic table, all of the various values for multiple parameters are stored in the table and then selectively retrieved using SQL statements or predefined view. The following example illustrates how to apply a generic table.

Case history: using a generic data table. This scenario involves the use of look-up lists in a data input form (see chapter *Designing the User Interface*). Briefly, the data input form requires that the user select a lithology type from a predefined list of acceptable input values. Normally, this involves programming the data values into the form or creating a single table that would include just the values for the list. An alternative is to use a generic data table. The following steps illustrate the procedure.

Step 1—Create the table. In this example, Access® was used to create a table consisting of five fields for each value stored. Figure 6–1 shows the basic structure of the completed table. The PARM_NAME field should be obvious. The next three fields allow for the data value in three different formats: text, numeric, and date (more parameters will be added to the table later).

Table: GENERIC_TABLE

Columns

Name	Type	Size
PARM_NAME	Text	50
PARM_VALUE_TEXT	Text	50
PARM_VALUE_NUM	Long Integer	4
PARM_VALUE_DATE	Date/Time	8
PARM_UNITS	Text	50

Fig. 6–1 Structure of a Generic Data Table. This example shows sample field sizes and data types for text, numeric and date fields. Other data types, field sizes, or numeric types may be necessary for a specific application

Step 2—Add data to the table. Next, the values that are to appear in the look-up box in the form are added to the table. Once this has been completed, the contents of the table look like Figure 6–2. Note that for this parameter no units are specified because it is text format data.

Designing the Database

PARM_NAME	PARM_VALUE_TEXT	PARM_VALUE_NUM	PARM_VALUE_DATE	PARM_UNITS
LITHOLOGY	Limestone	0		None
LITHOLOGY	Sandstone	0		None
LITHOLOGY	Dolomite	0		None
LITHOLOGY	Anhydrite	0		None
LITHOLOGY	Shale	0		None
LITHOLOGY	Granite	0		None
LITHOLOGY	Basalt	0		None
		0		

Fig. 6–2 Generic Data Table with Text Data Added. The example shows hypothetical text data added to the table. Note that the numeric field for the text data is all "0."

Step 3—Add other parameters. Next, a list of acceptable map scale values is added to the table that will be used in another look-up box (possibly in another form). Once these are added, the table contents will look like Figure 6–3.

PARM_NAME	PARM_VALUE_TEXT	PARM_VALUE_NUM	PARM_VALUE_DATE	PARM_UNITS
LITHOLOGY	Limestone	0		None
LITHOLOGY	Sandstone	0		None
LITHOLOGY	Dolomite	0		None
LITHOLOGY	Anhydrite	0		None
LITHOLOGY	Shale	0		None
LITHOLOGY	Granite	0		None
LITHOLOGY	Basalt	0		None
MAP_SCALE		1000		Scale
MAP_SCALE		2000		Scale
MAP_SCALE		10000		Scale
MAP_SCALE		50000		Scale
MAP_SCALE		100000		Scale
		0		

Fig. 6–3 Generic Data Table with Numeric and Text Data. The example shows the table after hypothetical text data and numeric data have been added. Note that the numeric field for the text data is all "0", and the text field for the numeric records is left blank.

Step 4—Retrieve values for a single parameter. To create the data list for the look-up lithology box, it is necessary to create a simple SQL query that will select only the data where the value in the PARM_NAME field is equal to LITHOLOGY. This can be done using a query-building wizard, or a SQL command can be written that retrieves the list. Either way, the results should look like Figure 6–4.

PARM_NAME	PARM_VALUE_TEXT
LITHOLOGY	Limestone
LITHOLOGY	Sandstone
LITHOLOGY	Dolomite
LITHOLOGY	Anhydrite
LITHOLOGY	Shale
LITHOLOGY	Granite
LITHOLOGY	Basalt

Fig. 6–4 SQL Results for Lithology Data from Generic Table. Note that only the text data (for lithology) is displayed.

The equivalent SQL statement that makes this retrieval (for those that are interested in such things) is as follows:

```
SELECT GENERIC_TABLE.PARM_NAME, GENERIC_TABLE.PARM_VALUE_TEXT
FROM GENERIC_TABLE
WHERE (((GENERIC_TABLE.PARM_NAME)="LITHOLOGY"));
```

To retrieve the list of acceptable map scales requires only a slight modification to the query to create the results shown in Figure 6–5. In the SQL statement, note that the only difference between this example and the preceding one is the WHERE clause, and the fact that we are now selecting the PARM_VALUE_NUM instead of the text field:

```
SELECT GENERIC_TABLE.PARM_NAME, GENERIC_TABLE.PARM_VALUE_NUM
FROM GENERIC_TABLE
WHERE (((GENERIC_TABLE.PARM_NAME)="MAP_SCALE"));
```

PARM_NAME	PARM_VALUE_NUM	PARM_UNITS
MAP_SCALE	1000	Scale
MAP_SCALE	2000	Scale
MAP_SCALE	10000	Scale
MAP_SCALE	50000	Scale
MAP_SCALE	100000	Scale
	0	

Fig. 6–5 SQL Results for Map Scale Data from Generic Data Table. Note that only the numeric map scale values are shown.

This example is extremely simplified, but it illustrates the potential for using generic data tables in a wide variety of applications. The possibilities are limited only by the user's imagination.

Now that the basic concepts of database design have been introduced, it is necessary to make sure that all the specific geotechnical data types and formats are included in the system. The next chapter builds on the earlier discussion of general data types and introduces the most common data types, format, and associated problems with geotechnical data.

7

Geotechnical Data

Many publications illustrate data management concepts with general examples from the world of accounting, finance, and human resources. While this is a good starting point, the range, complexity, and variability of geotechnical data make it a much more specialized area. This chapter introduces the most important types of geotechnical data, common data formats, and data specific to geoscience applications. Basic types of spatial (location), directional survey, depth-related, and time-related data are discussed in detail.

INTRODUCTION

Geotechnical data include all information needed to accurately describe features within an earth model. In a fundamental sense, there are only two types of geotechnical data: spatial data and time-related data. Spatial data include all information that relates to relative and absolute position or location within the earth model. This definition encompasses latitude, longitude, and elevation for virtually any type of geoscience information. Time-related data include not only calendar and clock time but also absolute time in terms of geological age.

Coordinate Data

Virtually all geotechnical data is spatial in nature. That is, earth-related data can always be expressed in terms of a position on or under the earth's surface. However, the representation of that position (which is always a relative position) depends on the frame of reference and the projection system used to model the earth's curved surface.

Latitude and longitude

The most common and widely applied reference system for positional data on the earth's surface is the latitude-longitude system of spatial coordinates. Latitude (the y value in a polar coordinate system) is the position on the earth's surface relative to the poles (north or south) and the equator. Latitude values can therefore range from 0° to 90°, either north or south. Longitude (the x value in a polar coordinate system) is the measurement (in degrees) east or west from the prime meridian (at 0°) to a maximum of 180°.

Latitude and longitude data should be stored in a geotechnical database in one of two ways: either as a single, numerical value in decimal degrees or as separate values for the degrees, minutes, and seconds (DMS) values.

Decimal degrees (DD) storage format. If coordinate (or any angular measurement data) data are stored as decimal degrees, some consideration should be give to the overall accuracy of the data relative to the numeric storage format. If the measured data accuracy requires six significant digits to preserve the accuracy, the data storage format should be an appropriate format to ensure that no rounding errors occur internally. Most internal data formats allow both 16- and 32-bit precision. For coordinate data, the highest precision available is generally the better choice.

While decimal degrees format is generally the best choice for storing coordinate data, it may not be the most practical for data entry. In these cases, a degree, minutes, and seconds (DMS) format may be a better choice. If possible, the user interface (see chapter *Designing the User Interface*) should provide the option to enter data in either format. Otherwise, the user will be forced to make tedious (and possibly incorrect) conversions before entering the data. To store latitude and longitude values in decimal degree (DD) format requires only two fields (in addition to the key field and other data fields) in a database table, as shown in Figure 7–1.

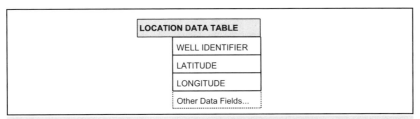

Fig. 7–1 Location Coordinates Stored as Latitude and Longitude; Angles in Decimal Degrees. This generic data table structure shows the minimum number of fields required to store coordinate data using this format.

Degrees, minutes, and seconds (DMS) format. The most common format for latitude and longitude data is still in degrees, minutes, and seconds. Unfortunately, this format complicates computational operations. If this is the format chosen for coordinate data storage, the DMS portions of the data must be stored as separate elements in the data table. In some cases, using this format for data entry can eliminate possible errors (e.g., degrees greater than 180, minutes or seconds greater than 60). If the original (observed) data are reported or measured in this format, then that should be the storage format in the database (see also the discussion in the preceding section regarding optional data input formats).

Storing coordinate data in DMS format requires six fields (in addition to the key field and other data fields) in a typical database table. Figure 7–2 shows a schematic table structure used to store coordinate data in DMS format.

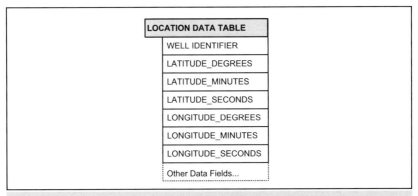

Fig. 7–2 Location Coordinates Stored as Latitude and Longitude; Angles in Degrees, Minutes, and Seconds. This generic data table structure shows the minimum number of fields required to store coordinate data using this format.

Regardless of the format chosen to store the coordinate data (and both formats may be used in the same database), there should be some facility to convert from one format to the other. Conversion utilities can be easily added to a form-based interface to perform these operations "on the fly." Pseudo-code for this conversion would take the form of Table 7–1.

Table 7–1 Conversion Pseudo-Code for Coordinate Formats

Conversion from DD to DMS format	Conversion from DMS to DD format
D = Integer(DD)	Result.DD = Degrees
M = (DD – D) * 60	Result.DD = Result.DD + Minutes / 0.60
S = ((DD – D) * 60) – M) * 60	Result.DD = Result.DD + Seconds / 0.0036

Universal transverse mercator projection method

The Universal Transverse Mercator (UTM) projection method is one of the most common methods of depicting the coordinates of a point on a curved surface. A complete discussion of the UTM projection method is beyond the scope of this book.

Using this projection method, any coordinate point (determined by latitude and longitude) can be converted to numeric x and y values that are relative to the same fixed reference as latitude and longitude. That is, a UTM-x value shows the location in meters east of a specific longitude (referred to as the *central meridian*). The UTM-y value is the location of the point in meters from the equator. During the computation, false northing and false easting values are added to ensure there are never any negative numbers. This serves a dual purpose, but mainly allows easy computational manipulation of UTM values. For example, the distance between any two points in a UTM coordinate plane only requires simple arithmetic, while determining the same information from latitude and longitude requires spherical trigonometry.

Meets and bounds

The system of meets and bounds is a specialized coordinate system used to reference polygonal areas on the earth's surface. If the database is intended to store this type of information (for the land department, as an example), some modifications to a conventional coordinate system may be required. Again, it is important to determine from the user ahead of time what the requirements are and the format in which the data will be stored.

Other coordinate systems

Various other projection systems locally provide the same type of coordinate information as the UTM system, such as the Lambert conformal projection and North American polyconic projection. The choice of projection system generally is not the decision of the data manager but is determined by local needs and user requirements. It is, however, very important to maintain information in the database that specifies what projection system is used, what the projection parameters are, and any other information needed to use the coordinates in a meaningful way.

DIRECTIONAL SURVEY DATA

Borehole deviation data, as determined from various directional survey methods, is used to describe the subsurface position of the wellbore in terms of x, y, and z (depth). The most common methods used to measure borehole deviation include gyroscopic surveys, electric multishot surveys, information from measurement while drilling (MWD), and (in most old wells) the single-shot or Totco® survey. All but the last of these surveys provide incremental measurement of the depth, borehole inclination, and borehole azimuth (the Totco® single-shot survey measures only the inclination at a specified depth, not the azimuth). To derive the actual x, y, and z values therefore requires that some sort of computation be made with the original observed data.

The basic concepts and abbreviations commonly used in directional drilling are shown schematically in Figure 7–3.

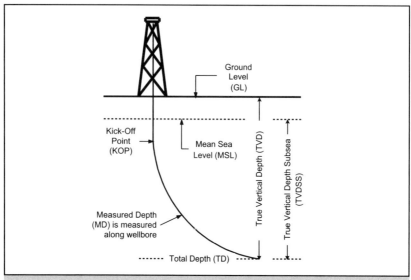

Fig. 7–3 Illustration of the most common directional drilling measurement relationships, nomenclature, and abbreviations.

Observed vs. computed data

As with virtually all geotechnical data, it is important to clearly distinguish between what is observed (*measured* data) and what is derived (*computed* data). Directional survey data are a particularly good example of why it is sometimes important to store the original observed data as well as the derived data in the same database.

Observed (measured) data. The basic borehole geometry or position is determined from the measured data using a variety of methodology. The basic recorded borehole geometry measurements are as follows.

- *Measured depth.* Normally, the directional survey is recorded at periodic increments, depending on the severity of the hole deviation. The most common measurement increment for vertical or moderately deviated wells is approximately 100 ft. Most wireline survey data are measured with an accuracy of 0.25 ft, so the deviation data should be stored with two significant digit precision.
- *Borehole inclination angle.* The angle that the borehole makes relative to a vertical reference is the borehole inclination angle. Most survey

methods measure this angle (in degrees) with three significant digit precision, so the data field used to store this value should be formatted accordingly. In the case of horizontal wellbores, it is possible to have boreholes that become negative at some point in the well trajectory; so this must be accounted for in the storage format and in the validation rules.

- *Borehole azimuth.* The direction in which the borehole deviates from the vertical is the borehole azimuth. Together, the inclination and azimuth can be viewed as a single vector. The azimuth value for most modern directional surveys is expressed in decimal degrees, with a range from 0° to 360°. Older directional surveys may be expressed in a quadrant format.

Quadrant and bearings. To nongeologists, the concept of quadrant angles may seem somewhat alien (and unfamiliar even to many younger geologists). The quadrant system is a widely used convention that divides the coordinate world into quadrants—northeast, southeast, southwest, and northwest. An angle with azimuth of 25° is expressed as N25E, while a direction of 190° is shown as S10E. While there is much practical value in the use of the quadrant system for field mapping, it becomes difficult to use when performing the complex mathematical algorithms used in deviation survey calculations.

Many older deviation data are in quadrant format, so the option to enter and edit data in this format may be necessary in the user interface (see chapter *Designing the User Interface*). If there is a possibility of encountering this type of data, it must also be initially accounted for in the database structure. As a minimum, the bearing value must be stored with a field for the north or south direction, the angle itself, and a field for the east or west direction.

Deviation data format variations. Through the history of the petroleum industry, there have been various standards and methods used to express deviation survey data. Because of these various formats and the desire to maintain the actual observed and measured data, it may be necessary to store much more than just the three basic data points (depth, inclination, and azimuth). While it is possible to make the conversions to a common format before storing the data, this is not advisable. Several possible data formats may be encountered.

Depth, inclination in decimal degrees (DD), and azimuth (in DD format) is the most common modern format for deviation survey data. This suite of data values only requires three fields in a data table (see Fig. 7–4).

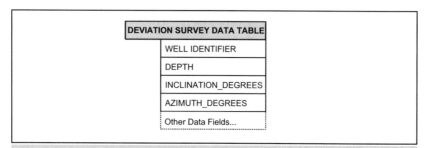

Fig. 7–4 Schematic data table structure showing the minimum number of fields needed to store deviation survey data using only depth, inclination, and azimuth, with angles in decimal degrees.

In some cases, the inclination and azimuth values are stored in degrees, minutes, and seconds (DMS) format. Representing the angular data in DMS format requires seven fields in the data table to store the information, as shown in Figure 7–5.

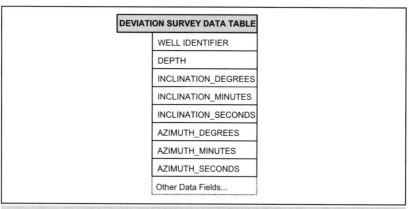

Fig. 7–5 Schematic data table structure showing the minimum number of fields needed to store deviation survey data (depth, inclination, and azimuth) where angles in are stored in degrees, minutes and seconds (DMS) format.

If the borehole orientation direction is shown as a quadrant bearing, a minimum of five fields are required in the data table to store the depth, inclination, and bearing data in the database, assuming all angles are decimal degrees. If all the angles are expressed in DMS format, the number of fields required jumps to nine, as shown in Figure 7–6.

DEVIATION SURVEY DATA TABLE
WELL IDENTIFIER
DEPTH
INCLINATION_DEGREES
INCLINATION_MINUTES
INCLINATION_SECONDS
BEARING_NORTHSOUTH
BEARING_EASTWEST
BEARING_DEGREES
BEARING_MINUTES
BEARING_SECONDS
Other Data Fields...

Fig. 7–6 Schematic data table structure showing the minimum number of fields needed to store deviation survey data where borehole orientation is expressed as a quadrant bearing and all angles are in degrees, minutes, and seconds.

Other directional survey data and information. In addition to the fundamental survey measurement data discussed in the preceding section, it is very important that other important information be captured and stored in the database.

- If logging tool information is available, the appropriate information about the individual tool should be recorded in the deviation database. This can be useful in the event that subsequent calibration problems or errors are found; the data associated with that tool can be corrected accordingly.

- In the same way, the logging unit information should be captured so that, if later problems arise (unit calibrations, etc.), the affected data recorded by that unit can be corrected.
- In multilateral wellbores, the directional survey is generally only measured in the lateral or sidetrack from the point of departure out of the main wellbore to the total depth of the sidetrack. Having the tie-on point (where the sidetrack connects to the main wellbore) allows reconstruction of the total survey from surface to total depth in all laterals and sidetracks. This is very important in ensuring each wellbore (lateral) has a complete survey that extends from surface to total depth.
- While most directional survey instruments are already calibrated to account for the magnetic declination (angular measurement between true north and magnetic north), this information should be stored in the database when available. As a quality control point, if the declination for a particular survey is incorrectly calibrated, the data can be adjusted accordingly.

Computed data

The most typical suite of computed data derived from the measured depth, inclination, and azimuth includes the true vertical depth (TVD), x and y offsets (relative to the surface point), dogleg severity, and offset along a vertical plane. In some cases, the original measured data are not available and only the computed depths and offsets are available. In these instances, it may be necessary to backcalculate the measured depth, inclination, and azimuth. When this has been done, the resulting data must be flagged as computed (and not measured), and the methodology used to make the calculations should be noted.

When a wellbore is deviated, the actual measured depth of a given point appears deeper than the actual or *true vertical depth* (TVD) below the surface. The TVD value is computed using the measured depth and inclination angle. This is a cumulative computation that is carried though the entire wellbore. For quality control purposes, scanning computed data for TVD points that are deeper than the corresponding measured depth points will flag many computational errors.

The position of a wellbore in the subsurface is measured relative to the surface coordinates of the well. This position is generally expressed as x offset and *y* offset or as offset in the north, south, east, or west, and is referred to as *relative offset data*. Assuming that the surface coordinates are correct, it is sometimes useful to store the actual UTM coordinates of the subsurface points as well (see *Storing Computed Results* section later in this chapter).

The *dogleg severity* is an angular measurement that shows the magnitude of the change in borehole direction (usually expressed in degrees per 100 ft). In most cases, this value is of no particular use to the geoscientist, but it can be very useful in the quality control of deviation survey calculations. Very large dogleg severity angles in unexpected places (e.g., in a straight portion of the well) can indicate data input errors or computational errors.

The *vertical offset* is the absolute offset of a subsurface point relative to the surface location of the well. Again, it is of limited direct use to the geoscientist but can be useful in screening data for potential data entry errors and computational errors.

Computational methods

The mathematical models that have been developed to model the position and geometry of the wellbore have evolved as a function of the methods available to make the computations. The most common of these are the angle averaging, tangential, radius of curvature, and minimum radius of curvature methods.

The earliest and mathematically simplest of deviation survey algorithms is the *angle averaging method*. This method was used when all calculations had to be done by hand using a slide rule and trigonometric tables. Because of the relative simplicity of the method, it is not considered accurate enough to be used today.

The *tangential method* was developed as computational devices become more sophisticated and widespread. It uses a more rigorous trigonometric solution. Again, with available modern computational methods and hardware, this method should not be used.

The *radius of curvature* and *minimum radius of curvature* methods (the latter tending to be the most accurate) apply spherical trigonometry algorithms to provide an accurate and reliable solution to the deviation data. Although several algorithms are available, the variations between them are minimal.

When dealing with legacy data, it is important to determine the computational methods used to make the calculations. In field development projects where drilling spans 20-30 years, there may be several vintages of methodology. This can yield inconsistent results that may have a significant impact on local structural interpretations. When this case is recognized, a complete recomputation of the original data using a minimum radius of curvature algorithm should be considered. This will yield a dataset that is internally consistent and will result in more consistent subsurface interpretations.

Storing computed results. As a general rule of good database design, derived or computed data should not be stored. Deviation data are one exception to this policy.

Because of the complexity of the calculations involved in deriving the actual subsurface location of a wellbore from measured data, it is sometimes more practical to store the derived data so that other (secondary) computations can be made more quickly. Otherwise, every time a subsurface point needs to be calculated or converted, the original measured data must be retrieved, a computational algorithm or function run, and the derived data returned. This can add undesired complexity and slow down the retrieval of data to an unacceptable level. If the computed data are stored in the database, however, storing all the required information about the computational method or algorithm is essential.

In addition, if the observed data are edited at any time, the editing action needs to trigger a manditory (and automatic) recomputation of the derived data as well.

Case history: consistent computational approaches. In many large, mature petroleum provinces there will be large volumes of legacy data spanning multiple decades of time. This case involved an international offshore operation that had a 30+ year drilling and production history that included hundreds of directionally drilled, platform-based wells. Although multiple wells were drilled from each surface location (platforms), the directional surveys conducted in the wellbores included virtually every historical survey method and computational algorithm. The resulting dataset was therefore inconsistent in several respects.

The solution in this case was to review all original observed coordinate and hole geometry data from all the wells. For directional surveys, this included the measured depth, hole inclination, and hole azimuth. Where multiple borehole surveys were available for a single wellbore, an industry-accepted quality and reliability ranking was established. Once all observed data were located, quality checked, and loaded to the database, an industry-standard computational algorithm was used to derive the borehole geometry (in this case, a minimum radius of curvature approach was used). This resulted in a dataset that was internally consistent, although it didn't always match historical derived data.

Once the new data were incorporated into the field interpretations, the resulting revisions in borehole geometry and horizon depths (and there were more than a few) created new opportunities in some areas but eliminated potential locations in others. At first, the new data were viewed with skepticism (especially in cases where drilling locations were eliminated in the revised interpretations). However, as new wells were drilled and the new interpretations were validated, the consistent data were embraced by the users.

Depth-Related Data

Most geotechnical data are derived from direct or indirect subsurface measurements and interpretations. As such, it is important to treat depth-related data as a very fundamental data type.

Depth reference data. Every depth point in the subsurface is referenced to a measured (or computed) point at or near the surface. It is therefore very important to always store the depth reference (TVD, TVD sub sea, TST, etc.) in the same record with depth-related data. In this way, there is no ambiguity as to the source of the depth measurement. In all cases, the actual measured depths should be stored because all other depths are derived from the observed measured depths.

Depth units. Surprisingly, there are no universal standards for depth measurement units. Depth values can be measured in either feet or meters, but vary from area to area and operator to operator. Thus, there are commonly variations in units convention even in the same field, depending on the well operator, the logging contractor, or the standards in place at the time the well was drilled. The depth units (meters or feet) should always be stored with the corresponding depth data to avoid any possible ambiguity or computational errors.

Depth range data. Geotechnical data involve many instances where data are referenced to a range of depths rather than a single depth point in the subsurface. One of the most common errors in both database design as well as database utilization is storing a range of depths as a text string. For example, a perforated interval can be shown by the text string 10545-10600 or by storing the numeric values of 10545 and 10600 in two separate numeric fields in the same table.

Whenever a depth range is specified, two data fields should always be included: one for minimum depth and the other for maximum depth. Data validation rules should make sure that the minimum depth is not greater than maximum (although it can be equal). If an exception field is included with the record, special cases can be identified (overturned, eroded, not deposited, repeated, etc.).

Stratigraphic Tops, Zones, and Markers

Although stratigraphic or formation tops and markers can be considered a special instance of depth-related data, these data form a fundamental part of any geotechnical database.

Formation tops are one of the most misapplied and misunderstood data types that face the data management professional. While the fundamental concepts are fairly well defined, how those concepts are translated from the real world of paper logs, colored pencils, and cryptic notations to the virtual world of an organized and consistent database becomes a substantial problem.

Every geoscientist has his or her own set of definitions and meanings for stratigraphic terms. Therefore, the first step is to clarify these definitions among all the users. Once a common basis of understanding has been established, the guidelines should be documented and all users should be informed of the standards.

Tops and markers

Formation tops and markers are the fundamental data types for identifying stratigraphic or seismic points in the subsurface. In most cases, a formation top or marker consists of a single depth point or a discrete surface. If the top is associated with a wellbore, the top is a point and should always be referenced to depth in that wellbore. A surface that represents a formation top can be depth-referenced or time-referenced (for seismic data).

Zones and layers

Reservoir or lithologic intervals are generally considered zones or layers. These zones have an upper and lower bounding surface defined by tops and/or markers. There are several ways to store zone or layer information in the database.

The single-record storage approach is generally used in spreadsheets and flat-file databases. In this format, the top and base of the pick or zone are stored together in the same record. For many applications, this approach is very practical and foolproof. Both of the bounding surfaces of the zone or formation are immediately available when processing the data or retrieving information, which can greatly simplify queries and reporting.

In a multiple-record storage format, the upper bounding surface of a zone is stored in one record and the lower bounding surface is stored in another record. Another table is used to relate the top and base of a particular zone or layer. While this provides a more normalized database and allows flexibility in dealing with the data, it makes retrieving the top and base a more difficult and indirect process involving complex relationships between several tables.

Stratigraphic exception codes

Stratigraphy is anything but routine and predictable. As such, there are a number of situations where normal and expected stratigraphic relationships do not exist. To deal with these problems and exceptions to normal situations requires the use of stratigraphic exception codes. These codes or flags are included in the database to provide valuable information regarding the stratigraphic relationships. Examples of stratigraphic exceptions include the following:

- Overturned beds due to structural folding
- Faulted intervals, including normal, reverse, and thrust faults
- Unconformities, nonconformities, and disconformities
- Nonpenetration. In cases where a well only partially penetrates a stratigraphic interval, it is helpful to flag these points with an exception code to prevent the partially penetrated sections from being included in isopach or other thickness calculations.

Stratigraphic nomenclature

The development of consistent stratigraphic nomenclature is very important to the success of a formation tops database. Wherever possible, stratigraphic nomenclature should adhere to standards. Once again, this is an area where it would be very useful to create an *ad hoc* committee to evaluate current nomenclature in use, anticipated expansion to the database, and formats that have already been adopted and are in current use.

A suggested formation naming convention involves the use of a hierarchical approach to nomenclature. The name should include an abbreviated code for the formation name as a prefix and a suffix that allows a high degree of subdivision to allow for refined layer definitions for reservoir modeling. For example, a series of formation names might include the following:

ARCH0001
BRAZ0010
CRZE0025

In this example, the three formation names uniquely identify the formation and the numeric subdivision of that formation. Note also that limiting the complete name to eight characters or less provides a backward compatibility with those (hopefully rare) situations where an eight-character file-name limitation applies. For example, if formation characteristics are exported from a database into a file, it is convenient to use the formation name as the file name. By limiting the file name to only eight characters, the file is completely compatible with any application that imposes a file-name length limit.

Tops and bases. Two groups of geoscientists routinely deal with formation tops. One group advocates the use of a top pick for any given formation, with the base assumed to be the top of the unit immediately underlying it stratigraphically. The other group prefers to provide a top and a base for each zone or layer.

The disadvantages of the first approach should be obvious. First, if the underlying formation is variable (due to unconformities or faulting), then the default base (i.e., top of the underlying formation) will invariably be incorrect. By providing a top and base, however, this situation is avoided. This approach even allows for a situation where the base of one layer or zone is not the same as the underlying formation or zone.

The obvious exceptions to this rule are horizons, markers, and fluid contacts. These, by definition, have no thickness. In these cases, the marker top and base are the same value.

Future expansion capability. The use of a numeric suffix in the formation name provides several advantages. Selecting the initial numeric values with enough separation between incremental layers allows additional subdivisions of layers within the zone that may be used for simulation and/or modeling. The ordered numeric suffix approach also allows the layers to be ordered or sorted in their proper stratigraphic sequence.

Ownership issues. If two geologists pick the same formation in the same well, there will likely be three different values (assuming that one of them cannot decide between two picks). With this in mind, the database should be designed to accommodate some method of identifying the owner or originator of the stratigraphic tops data. This could be the user's initials, computer identification (ID) code, or some other simple code. Identifying any data with a user ID code of some type allows prioritization of data selection, depending on the reliability of an individual user, how outdated the information is, or any other combination of selection criteria.

Top criteria. A commonly overlooked aspect of stratigraphic tops databases is a method for determining the geoscientist's justification or rationale for picking a particular top. In the simplest form, this can be handled with a memo format field that allows the user to write a brief description of the criteria used for picking the top, the well type, the log curves used, or other information that will allow others to evaluate the pick process. A more sophisticated approach would include an image of the section of logs used, or other key reference data that would help clarify these criteria.

TIME-RELATED DATA

In the geophysical world, subsurface data are most commonly referenced in terms of time rather than depth. A discussion of the details of storing seismic data is beyond the scope of this text, but some of the main points relative to data management are considered here. Time-related data also include geologic age data and paleontological data that provide both absolute and relative stratigraphic time.

Geophysical data

Geophysical data management has become increasingly more complex as more sophisticated geophysical acquisition methods and computing power have increased data volumes by orders of magnitude. Fundamentally, we are dealing with the same types of data—just much larger volumes and a wider array of seismic attributes. This text cannot provide a complete discussion of the complex issue of geophysical data management; rather, the focus will be on general data types and considerations when dealing with this type of data.

Navigation data (shotpoint and line information). One type of geophysical data has changed little through the years: line navigation data. Virtually all geophysical data involve the use of point sources of energy (shotpoints) and microphones (receivers). Since each source and receiver point is spatial data, the x and y coordinates and elevation (z value) need to be maintained in the database. In this regard, the spatial data points in a seismic dataset need to follow the same general rules as all other spatial data (such as wells). In addition to the spatial coordinates of shot and receiver points, several other parameters need to be addressed in the database.

The *line naming convention* used for 2-D seismic lines rarely adheres to any set conventions. Since this is the key index for the line, standards similar to those developed for wells should be adopted. If it is not possible to create a new naming convention, a unique line index code should be created in the database to provide a consistent and unique reference to the line. Using a naming convention similar to the procedures for wells and formation tops is highly recommended for line names.

For 2-D seismic lines, it is sometimes useful to store the *coordinates of the line endpoints*. This is especially helpful if the line locations need to be plotted quickly on a map-based interface. While this convention does not provide shotpoint details (or allow for sinuous lines), it is a rapid way of depicting line locations on a map. In addition, the line endpoints can be used in a query of geographical boundaries to locate lines that fall within a selected area. For example, to perform a search query on line endpoints, a rectangular area using geographic coordinates can be specified to retrieve any lines that fall within the rectangle.

It is helpful to store basic *survey information* in the database to simplify retrieving metadata about the line. Such information should include line length, shotpoint spacing, number of shots, energy source, and source depth.

For 3-D surveys, additional information on bin size, inline-crossline spacing, and survey boundary coordinates, etc., should be included.

Seismic trace data. The actual seismic trace data is a much more difficult data type to deal with in a traditional data management structure. Most seismic data are recorded and stored in SEG-Y format, an industry standard ASCII text format that is almost universally imported and exported by major interpretive applications (see Appendix A for further definition of these formats). Because of the nature of seismic trace data, it is usually better to leave these data in an SEG-Y file format and then store either the file itself (in a binary field in the database) or a file pointer that identifies the location of that file. Both methods are acceptable, although storing the file itself eliminates the problem of maintaining a separate data storage server for the actual files. This eliminates the additional problem of maintaining the file directory structure, since the reference pointers in the database could be invalidated if the files are moved to a new location in the system.

Statics and velocity data are specialized data types that are generally difficult to manage in a conventional database. Images of the velocity profile data can be easily stored in binary large-object (BLOB) fields in the database, and statics information can be stored in a conventional table structure. In most cases, this sort of information can be managed by the interpretive application used for processing and/or interpretive work.

Checkshot survey data can be managed in much the same way that formation tops or deviation data are handled. Since checkshot surveys consist of survey data at discreet but variable depth points, they should be stored in a table of depth-related data. In addition to being used for seismic velocity analysis and the creation of synthetic seismic traces, checkshot data can be very beneficial in converting other subsurface information from depth to time. In some interpretive applications, there are no built-in functions for converting well traces and formation tops (measured in depth) to two-way time so that the wellbore and tops information can be displayed with geophysical data in time. This may require the use of a programmatic approach with a custom procedure or function.

There are two basic components of a *vertical seismic profile* (VSP): basic checkshot data and waveform data. The checkshot component of the VSP can be handled in the same way that a conventional checkshot is managed. The waveform portion of the survey is fundamentally seismic trace data, and can be managed as SEG-Y data.

Geological age data

While not the same sort of time-related data as geophysical data, geological age data are nonetheless time oriented. Instead of milliseconds, geological time is measured in millions of years.

The use of carbon-14, fission track, radiometric, and other *age-dating techniques* can provide valuable stratigraphic age data for the geoscientist. Absolute geological age is, of course, a single time measurement at a discreet depth point in the subsurface (or at a specific set of coordinates in an outcrop). While subsurface depth-oriented age data are straightforward from a data management standpoint, the issue of age dating from surface samples requires a more indirect approach. In this regard, it is necessary to relate the age date information to stratigraphic position using a table to store stratigraphic column information . The absolute age is related to a certain position in a stratigraphic section, which in turn is related to a specific depth in the subsurface through a formation tops table.

Geological time is a very imprecise type of data, and in most instances, there can only be *age range information*—an estimate of the minimum and maximum age provided. This is particularly true of paleontological data, where particular faunal assemblages are recognized that have a generally accepted range of ages. Like absolute age information, it is possible to relate time to a specific formation or even depth point, but it requires a more complicated relation between the age data, formation names, and subsurface tops. In many respects, managing paleontological data is unique and may require specialized application software rather than a master database.

The most useful data management approach for handling geological age data involves one or more tables of *stratigraphic information*. The basic structure of such tables includes general information about the stratigraphy with corresponding related absolute ages and age range information.

LOG AND BOREHOLE DATA

Wireline log data are one of the most fundamental types of geotechnical data, and they are used by geologists, geophysicists, petrophysicists, and engineers to determine subsurface characteristics and rock properties. Because of the wide use and primary importance of this data type, a great deal of attention should be given to designing the portion of the database that handles the log data. Many software and logging services companies have developed specialized log data handling systems. In most cases, these application systems work in conjunction with a specialized interpretation system and/or data storage and retrieval system. However, some of these systems are designed to be compatible only with certain software products.

When considering a log data storage and retrieval system, many critical factors must be taken into account. As with any software application, the end use of the data must be considered the primary factor driving the selection. If the primary end use of the data is geological (i.e., stratigraphic interpretation), then a system should be selected that provides maximum flexibility in the storage, transfer, and display of the log data. If the end use of the data is petrophysical in nature, analytical functionality becomes the most important consideration. As always, the end user or user group should have a significant amount of input into defining the selection criteria, testing the product for suitability, and assisting with deploying the product.

General organization of log data

Borehole logging data consist of two fundamental types: header information and curve or trace information. As with traditional paper logs, the log header provides a wealth of information (sometimes more than the user needs) about the location, depths, times, drilling fluids, logging parameters, tool specifications, and calibration. The actual log data consist of multiple channels or suites of continuous, vector-type curve data. Each log curve represents a particular type of subsurface measurement.

Header information. Like the header (and footer) sections of a traditional printed paper log, digital log header information must contain all pertinent information about a particular logging run, log suite, or log processing sequence.

Printed log headers generally have *local coordinate and elevation information* for the wellbore. Digital log datasets are inconsistent in this regard and cannot be assumed to have all of the necessary spatial data. Furthermore, the coordinate data on the log header are generally placed there by the logging engineer at some point during the logging job. The number of places where that information can be corrupted makes the log header location information the last place one should consider the definitive source of spatial data on a well. Surprisingly, many geoscientists will not accept location data unless it matches the log header (even when a predrill and postdrill GPS survey of the wellhead is available).

General logging parameters on a printed log header provide information about mud resistivity, maximum recorded temperature, time since circulation stopped, mud filtrate resistivity, mud type, and various other run-specific logging parameters. These parameters need to be maintained in the database so that log data from a particular logging run can access the appropriate header parameters.

On a printed log, the *curve header information* is shown above the actual curve trace data, which include the curve name or mnemonic code, curve scale, and units. In many digital log storage and transfer formats, much of this key information is lost (this is especially true of curve scale values). Therefore, it is important to maintain this information in the database so it can be properly transferred to other applications when needed.

The actual *log curve traces* are generally considered vector data rather than point data (even though the data are sampled at regular depth intervals within the wellbore).

Log data storage and transfer

Log curve data can be stored and transferred between databases and applications in several formats. The storage format can be independent of the transfer format, depending on the design of the database and import/export capabilities of the DBMS and interpretive applications. There are two types of data formats for log data: ASCII text (commonly the Log ASCII Standard, or LAS, as discussed in the next section) and binary.

LAS format. The Log ASCII Standard (LAS) format was developed by the Canadian Well Logging Society (CWLS) in the 1980s as a simple, efficient method of storing and transferring log data on floppy disks. The original LAS 1.2 format was quickly adopted by the petroleum industry by both end users and logging contractors. This standard format continued to evolve with the changing needs and requirements of users and contractors and is now considered a de facto standard for nonbinary log data formats.

LAS 1.2 consists of a log header section, a logging parameter section, a curve header section, and a log data section (Fig 7–7). The entire file is plain ASCII standard text and can be read by virtually any modern log interpretation application. The main limitation to the original version was that the number of curves that could be included in a file was limited to the width of the ASCII record length (256 characters). For a small number of curves, this was not a problem, but the format became more difficult to use as log suites became more complex and curves became more numerous.

```
EXAMPLE #1 - THE LAS STANDARD IN UNWRAPPED MODE

~VERSION INFORMATION
VERS.                    2.0      : CWLS LOG ASCII STANDARD -VERSION 2.0
WRAP.                    NO       : ONE LINE PER DEPTH STEP
~WELL INFORMATION
#MNEM.UNIT              DATA         DESCRIPTION OF MNEMONIC
#----- -----          ----------   -------------------------
STRT .M               1670.0000     :START DEPTH
STOP .M               1669.7500     :STOP DEPTH
STEP .M                 -0.1250     :STEP
NULL .                -999.25       :NULL VALUE
COMP .      ANY OIL COMPANY INC.    :COMPANY
WELL .      ANY ET AL 12-34-12-34   :WELL
FLD  .      WILDCAT                 :FIELD
LOC  .      12-34-12-34W5           :LOCATION
PROV .      ALBERTA                 :PROVINCE
SRVC .      ANY LOGGING COMPANY INC.:SERVICE COMPANY
DATE .      13-DEC-86               :LOG DATE
UWI  .      100123401234W500        :UNIQUE WELL ID
~CURVE INFORMATION
#MNEM.UNIT       API CODES       CURVE DESCRIPTION
#----------      ---------       -------------------------
DEPT .M                          :  1   DEPTH
DT   .US/M       60 520 32 00    :  2   SONIC TRANSIT TIME
RHOB .K/M3       45 350 01 00    :  3   BULK DENSITY
NPHI .V/V        42 890 00 00    :  4   NEUTRON POROSITY
SFLU .OHMM       07 220 04 00    :  5   RXORESISTIVITY
SFLA .OHMM       07 222 01 00    :  6   SHALLOW RESISTIVITY
ILM  .OHMM       07 120 44 00    :  7   MEDIUM RESISTIVITY
ILD  .OHMM       07 120 46 00    :  8   DEEP RESISTIVITY
~PARAMETER INFORMATION
#MNEM.UNIT          VALUE            DESCRIPTION
#---------       ---------        ---------------------
MUD  .           GEL CHEM          :   MUD TYPE
BHT  .DEGC       35.5000           :   BOTTOM HOLE TEMPERATURE
BS   .MM         200.0000          :   BIT SIZE
FD   .K/M3       1000.0000         :   FLUID DENSITY
MATR .           SAND              :   NEUTRON MATRIX
MDEN .           2710.0000         :   LOGGING MATRIX DENSITY
RMF  .OHMM       0.2160            :   MUD FILTRATE RESISTIVITY
DFD  .K/M3       1525.0000         :   DRILL FLUID DENSITY
~OTHER
    Note: The logging tools became stuck at 625 metres
    causing the data between 625 metres and 615 metres to be
    invalid.
~A  DEPTH       DT      RHOB        NPHI       SFLU      SFLA      ILM       ILD
1670.000    123.450   2550.000     0.450    123.450   123.450   110.200   105.600
1669.875    123.450   2550.000     0.450    123.450   123.450   110.200   105.600
1669.750    123.450   2550.000     0.450    123.450   123.450   110.200   105.600
```

Fig. 7–7 Example open-hole log data using the Log ASCII Standard (LAS) version 1.2. Note that there is one line of log data for each depth point when this format is used.

Version 2.0 of the LAS standard improved on the original format by allowing line wrapping. That is, instead of all the curve data values for a particular depth being contained on a single line in the file, the line could wrap to the next line in the file. This revision produced a format that increased the quantity of curves to a virtually unlimited number (Fig 7–8).

The most recent version of the LAS standard, 3.0, has added features to the line-wrapped version 2.0 that greatly expand the number of log header parameters that can be included in the file. Additional information on CWLS and the LAS format can be found in Appendix A.

One drawback to LAS is the speed with which individual points in the dataset can be loaded or accessed by an interpretive application. For this reason, most software provides a means of loading data from LAS format files and then stores the data internally in a binary format.

A complete discussion of the LAS 2.0 and 3.0 standards, with format details and example datasets, can be found at the CWLS Web site (see appendix A for additional information and Web site links).

Binary formats. The various binary formats used to store and manage log data provide a significant advantage over the LAS standard in how fast log data can be accessed by the computer system. Petrophysical and geoscience software applications generally store log data internally, usually in some sort of proprietary format. This allows the interpretive application rapid access to the data for computations and display purposes. However, several industry-standard binary data formats are used to store and transfer log data directly in a binary format.

The Log Information Standard (LIS) was developed early in the history of digital log processing as an efficient binary format for recording and transmitting log curve data. Since the format is binary, special software utilities are required to read or write the data. However, the individual LIS-format files can be stored and managed without resorting to these additional utilities.

The Digital Log Information Standard (DLIS) is an enhancement of the standard LIS format and creates a more highly compressed format. The DLIS format also allows for the storage of the large volumes of array-type data recorded by specialized array tools and waveform logging devices.

```
              EXAMPLE #3 - ILLUSTRATING THE WRAPPED
                    VERSION OF THE LAS STANDARD

~VERSION INFORMATION
VERS .                      2.0    :   CWLS log ASCII Standard -VERSION 2.0
WRAP.                       YES    :   Multiple lines per depth step
~WELL INFORMATION
#MNEM.UNIT           DATA            DESCRIPTION OF MNEMONIC
#----- -----        -------         ---------------------------
STRT .M              910.0000       :START DEPTH
STOP .M              909.5000       :STOP DEPTH
STEP .M               -0.1250       :STEP
NULL .                -999.25       :NULL VALUE
COMP .     ANY OIL COMPANY INC.     :COMPANY
WELL .     ANY ET AL 12-34-12-34    :WELL
FLD  .     WILDCAT                  :FIELD
LOC  .     12-34-12-34W5            :LOCATION
PROV .     ALBERTA                  :PROVINCE
SRVC .     ANY LOGGING COMPANY INC. :SERVICE COMPANY
SON  .     142085                   :SERVICE ORDER NUMBER
DATE .     13-DEC-86                :LOG DATE
UWI  .     100123401234W500         :UNIQUE WELL ID
~CURVE INFORMATION
#MNEM.UNIT                     CURVE DESCRIPTION
#---------                     --------------------
DEPT .M                        :   Depth
DT   .US/M                     :   1 Sonic Travel Time
RHOB .K/M                      :   2 Density-Bulk Density
.
. Some curves omitted from listing
.
FBH  .                         : 33 Flag -Bad Hole
FHCC .                         : 34 Flag -HC Correction
LSWB .                         : 35 Flag -Limit SWB
~A Log data section
910.000000
-999.2500  2692.7075      0.3140     19.4086    19.4086    13.1709   12.2681
  -1.5011    96.5306    204.7177     30.5822  -999.2500  -999.2500    3.2515
-999.2500     4.7177   3025.0264   3025.0264    -1.5010    93.1378    0.1641
   0.0101     0.1641      0.3140      0.1641    11.1397     0.3304    0.9529
   0.0000     0.1564      0.0000     11.1397     0.0000     0.0000    0.0000
909.875000
-999.2500  2712.6460      0.2886     23.3987    23.3987    13.6129   12.4744
  -1.4720    90.2803    203.1093     18.7566  -999.2500  -999.2500    3.7058
-999.2500     3.1093   3004.6050   3004.6050    -1.4720    86.9078    0.1456
  -0.0015     0.1456      0.2886      0.1456    14.1428     0.2646    1.0000
   0.0000     0.1456      0.0000     14.1428     0.0000     0.0000    0.0000
909.750000
-999.2500  2692.8137      0.2730     22.5909    22.5909    13.6821   12.6146
  -1.4804    89.8492    201.9287      3.1551  -999.2500  -999.2500    4.3124
-999.2500     1.9287   2976.4451   2976.4451    -1.4804    86.3465    0.1435
   0.0101     0.1435      0.2730      0.1435    14.5674     0.2598    1.0000
   0.0000     0.1435      0.0000     14.5674     0.0000     0.0000    0.0000
909.625000
```

Fig. 7–8 Example open-hole log data using the Log ASCII Standard (LAS) version 2.0. Note that log data records (lines in the file) for each depth point are wrapped when this format is used. This allows a virtually unlimited number of curves to be included in a single file.

Throughout the history of the logging industry, various other proprietary data formats have evolved, many designed and used by specific companies as a corporate standard. Most of these are gradually being displaced in favor of the more widely adopted binary and nonbinary formats discussed here. During the development of a DBMS project, it is important to know whether any of these specialized formats will be encountered as input or required for output so the database and data delivery tools can be modified accordingly.

Data editing considerations

When planning and executing a data management project, the issues relating to log data editing will become very complex and hotly debated. While not directly the responsibility of the data manager, a fundamental understanding of the process of log data editing is important because this process impacts or is impacted by the database issues at several points. The major steps in the acquisition and editing of log data include the following.

Original log data. It is essential that the original, as-recorded log data be stored in the database and identified as the original logs. While many debate the value of this statement, once the original data are edited and stored in place of the original data, there is no turning back. If an editing error is later discovered and the original log data are not available, it is sometimes impossible to recover the original data. Of course, the original data must be identified in the database as original, and the following basic information must be recorded in the database.

The *unique well identifier*, or UWI, for the wellbore, can be a purely numeric value or it can be a more human readable format. One approach is to combine a short text string that identifies the field, well number, and borehole number in a standardized format. For example, the general format XXXX_nnn_nn allows four characters for the field name (XXXX), three digits for the well number (nnn), and two digits for the borehole number (nn). Additional identifiers can be included, but this format can serve as a basic template.

The unique *borehole identifier* is used to relate the logging data to a particular borehole in the well (assuming there are laterals or sidetracks). If the borehole number is included in the unique well identifier, this is not necessary.

Logging run information is required so the data from a particular logging run can be related to the appropriate logging parameters from the log header data.

In many cases, the log data for a particular curve trace or tool will not extend over the entire interval of that logging run. It is very important to know where there are real data for that curve, such as the *minimum and maximum logged depths*.

Borehole environmental corrections. The petrophysical corrections needed to account for pressure, temperature, borehole size, and other environmental factors need to be available on a run-by-run basis. This is necessary because the header parameters for each logging run vary slightly (or a great deal), and it is important to make these corrections prior to splicing and editing. The environmentally corrected data should also be stored in the database, along with any comments and correction parameters made by the petrophysicist. This will be helpful in the event there is ever a need to revise the environmental corrections.

Splicing. In all but the simplest of wells, there will probably be more than one suite of logs run at each casing point in the drilling operation (ignoring the through-casing logging data). The goal for the user, however, is to have a continuous log curve trace for each log rather than individual pieces. The first step in the splicing process is to splice each run of each log into a continuous curve trace that extends over the entire data interval (which may not be the entire well). The continuous logs should be stored and identified as spliced logs in the database. In some cases, this version of the log data is used as the working set for most geoscientists.

Editing. After borehole corrections and splicing is completed, the petrophysicist makes various editing corrections using a variety of techniques to eliminate shifted data, measurement drift, data noise, etc. The edited logs are then identified as such and stored in the database.

Normalization. The final and most often ignored step in the process is the normalization of log responses. This is a time-consuming and tedious job and is sometimes not possible with log data from remote areas. If normalization is performed, the normalized data should be stored and identified in the database. The general process of log data normalization is discussed in the *Data Validation, Editing, and Quality Control* chapter.

Problems of log data management

The management of log data is no trivial task. The multiple generations and vintages of raw and edited data (not to mention the computed data) create an enormous data management problem that results in special considerations.

Original, raw log data represent the only observed measurements and as such should never be modified. Any editing modifications to the raw log data must be attributed to a data owner. Generally, the log analyst or technician performs the editing, normalization, or computation based on the raw log data. Part of the log data management system must include a method of tracking log data ownership through the various processing and interpretation steps.

Log interpretation generally requires numerous processing and interpretation steps. As a result, there will generally be several vintages of data and several different interpretations that incorporate different parameters or algorithms. The log data management system must include some method of setting access priorities or preferences so the user can select which data versions will be available and in what order they will be selected.

Other wireline data

There are a wide variety of specialized nuclear, acoustic, and resistivity tools, but most of these can be managed as conventional wireline data (with the exception of array tools and cement bond logs).

Wireline formation testing tools are used to conduct pressure tests and/or collect fluid samples from one or more intervals in a borehole. The data acquired from a formation testing device are therefore from discrete depth points, but include continuous pressure measurements as a function of time. Most log data management systems do not directly handle formation testing data, and it may be necessary to store the data in a custom application database that is specifically designed to handle it. It is possible to store and manage formation testing data in a relational database, but custom extensions to commercial geotechnical data models may be necessary. In this case, there are clear advantages to leaving the data in an application-specific database and simply storing the interpreted pressure data in the master database.

Vertical seismic profiles (VSPs) and conventional seismic checkshot data can be managed to a certain extent by conventional relational database models. The velocity data at discrete depth points (from both VSP and conventional checkshots) are relatively straightforward in terms of a relational database. The full waveform data from the VSP, however, are more suitable stored in SEG-Y or other conventional geophysical data format. In any case, it is critical that the proper reference elevations, survey parameters, and other data are stored with the checkshot information.

PETROPHYSICAL DATA

Petrophysical data can be derived from direct measurements and observations (as in the case of conventional and sidewall cores) as well as indirect measurements of electrical, nuclear, or acoustic properties in the case of wireline logs.

Petrophysical data from cores

Core data, both conventional and sidewall, provide the only direct physical ground truth of subsurface geology and rock properties. As such, core data are an integral part of the database and are used (or should be used) to calibrate and normalize all other indirect measurements.

The general information about the core data is required to locate and identify the information accurately. This includes the top and base of the cored interval, coring method (conventional, sidewall, rotary sidewall), the percent recovery, core diameter, and all basic information required to link the core data to the well location information. If the core is obtained in a sidetrack or multilateral borehole, the actual borehole information is also mandatory to link the core with the correct sidetrack in the well.

Porosity and permeability are the two most common measurements gathered from core analyses. However, it is also important to capture the type of porosity analysis method used, vertical permeability, and horizontal permeability. If the analysis was conducted at simulated reservoir pressure, a separate set of analytical data must be recorded for each pressure condition.

Clay volume determination is included in some core analyses. In addition to the actual volume measurement, the analytical parameters should be recorded.

Special core analyses (SCAL) yield data that are somewhat more difficult to manage in a conventional database structure. Mercury injection capillary pressure data include both sequential data (which can be stored in the database as discrete values), and graphical results. Ideally, the database should include some means of storing an image file of the graphical results in a binary data format field.

Indirect (computed) petrophysical data

Most geoscientists are acutely aware that wireline logging devices do not directly measure porosity, fluid saturation, or clay volume. All of these parameters are computed from direct acoustic, nuclear, and resistivity measurements recorded by the logging tools. These parameters are considered indirect data because they are computed from direct measurements of other parameters.

The single most important petrophysical parameter is porosity. In addition to the actual porosity value, the database should store information regarding the source of porosity (tool or tools used) and the algorithm used to compute the porosity. The data must also adhere to internal standards regarding the data format (decimal percent or percent).

If there are pore spaces in the rock, they must contain some volume of fluids (liquids and/or gases). As with porosity, an internally consistent data format must be used to store fluid saturations. Since hundreds of algorithms can be used to compute fluid saturations, storing a reference to the methodology used should also be mandatory.

The volume and type of clay in the reservoir can also be computed from log data using a variety of logging tools and computational algorithms. As with fluid saturations, dozens of clay volume algorithms are available. Storing a reference to the algorithm (or the equation itself) is important to be able to interpret the results. Again, a consistent data format should be maintained.

Various combinations of acoustic, nuclear, and resistivity logging tools can be used to estimate mineralogical volumes at any given point in the wellbore. The algorithms and methodology used are very sophisticated, employing both deterministic and stochastic modeling. Obviously, the methodology and logging tools that compute lithology should be included in the database. Storing the actual mineral volumes presents a somewhat unique problem of data storage. Because there can be a variable number of minerals, it is quite difficult to create a standard data model that can account for all possible mineral types without creating an enormous and complex data table, much of which will be empty depending on the reservoir. A possible solution to this problem is to use a modified version of the generic data table, which could include a well identification (ID) reference and a depth reference in addition to the data fields (see discussion on generic data tables in the preceding chapter).

Derived petrophysical data

Beyond the observed wellbore data and the computed data that are derived from these measurements, there is another class of petrophysical data. Information that is computed from multiple indirect calculated values is considered derived.

The determination of permeability from wireline data has been the ultimate goal of wireline logging since the days of the first wireline surveys. After all, a reservoir can have porosity and hydrocarbon saturation; but without permeability, it cannot be recovered economically. Permeability data determined from log data require that the algorithm, curves used, and other parameters be stored in the database.

Individual organizations have differing criteria by which net reservoir rock is determined, but these important data values are essential information and should be included in the database even though they are considered derived data. Some provision in the data model should be made, however, to include the criteria by which the net reservoir rock is computed. Some of these criteria include the following:

- The log curves used to determine net pay should be identified, including version number, run number, and any other identifying parameters.
- Since multiple curves are normally used to determine net reservoir, the minimum and/or maximum values for each curve used to set the limits of net reservoir should be recorded.
- Identifying information about the user making the net reservoir determination, including the date.
- A memo-format field should be included that allows the user or subsequent interpreter to make free-form text comments about the curve data, criteria, or other pertinent information that will be useful in understanding the thought process used to determine net reservoir.

The portion of the reservoir that is considered productive is unique, in that it is determined from a combination of both indirect (calculated) and derived data (such as permeability and net reservoir). The same criteria as defined for net reservoir values (log criteria, cut-off criteria, user information, and remarks) should be stored for net pay values, with at least one additional piece of information.

Since net pay depends on other variables (such as permeability and net reservoir), any other criteria used to determine it should be stored in the database. Depending on the maturity of the database, it may be possible to use a predefined list of criteria used (based on the existing contents of the database), or it may be necessary to store this information in a free-form memo field.

Data management problems of petrophysical data

In summary, petrophysical data present unique challenges to the data manager. Most of these problems can be avoided through careful planning and by maintaining good communications with the petrophysicists and log analysts.

Log analysis is both science and art, and it can be very subjective. As such, the individuals who create petrophysical interpretations provide very individual interpretations of the data. Because of limited log analysis resources in most companies, it is very important to identify the lineage of any interpretations and calculations. In this way, if changes in processing parameters or interpreters occur, it is a simple matter to identify each user's interpretations.

Log data processing has become very sophisticated since the old days of paper logs, slide rules, and chartbook solutions. With the advent of complex array and waveform logging tools, log data processing has become nearly as complex as seismic data. Therefore, it is important to create some means of storing all necessary information about each step in the processing stream.

If there is more than one geoscientist or log analyst involved in interpreting log data, there will be more than one interpretation. Each user or interpreter should have the option of storing a unique interpretation version in the database. Therefore, the database must be able to store user information, interpretation version, and usage priority. The usage priority allows other interpreters to select the priority of versions or interpretation sets based on user information.

Spatial Data and GIS Systems

Virtually all geotechnical data have spatial coordinates because they are all earth-related data. That is, any earth-related data can be described in terms of 3-D space (x, y, and z). With the increasing use of Geographic Information Systems (GIS), geoscientists are finally realizing a long-awaited dream. The ability to browse geotechnical data on a map-based interface and to interactively select specific items on the map for more detailed information has been in a long time coming but is still not widely adopted for many reasons discussed in this section. This ability to "click" on a feature displayed in a GIS interface and display information about that feature is referred to as "drilling down." The ability to browse a map-based interface and "drill down" on a particular data item is entirely dependent on how (and if) the underlying data have spatial references (i.e., the 3-D coordinates needed to reference it to the map) and appropriate links to the actual data tables.

The most popular and widespread GIS application suite in use today is the ArcView® product suite developed by ESRI. A number of vendors still provide GIS functionality in their products with proprietary map engines, but the more versatile and powerful features of the ArcView® GIS browser are gradually replacing these. The major advantage of standardizing with a commercial product such as ArcView® is that support, maintenance, and continued development are supplied by the vendor. ArcView® also provides excellent tools for developing custom extensions and functions, making the browser truly a custom product tailored for the specific needs of the company or individual.

Geotechnical data and GIS applications

Every geoscientist has seen numerous vendor (or even internal) presentations on the power and potential of using GIS applications as a window into the geotechnical database. No one can dispute the power and potential of this technology or the need for this type of interface. However, the widespread application of GIS methods is still limited in most cases by the availability of detailed spatial data and the metadata needed to access it.

Spatial metadata. Most geoscientists poorly understand the study of spatial metadata, although it is a key element in the successful application and implementation of GIS technology. Metadata is the information that describes the spatial data, not the spatial data itself. For example, the actual spatial data for a map would include the coordinates of the corner points of the map, but the metadata would describe the origin of the map, the version, source of data, projection system used, etc. In its simplest form, spatial metadata requirements are straightforward; but to fully describe a geotechnical dataset in terms of spatial parameters is very complex. A complete discussion of spatial metadata requirements can be found at the Federal Geographic Data Committee Web site (see Appendix A for further information about the FGDC).

Data formatting issues. Most GIS browser interfaces require the data to be in very specific formats, which may not be available in existing export functions. Spatial data are normally represented on a GIS interface as "shapefiles." Shapefiles are a series of points that identify the outline of the spatial data. While it is possible (and preferable) to extract data from the database on demand using direct database links (such as ODBC), it may be necessary to store the shapefiles that represent the data in a new or extended table. If this approach is used, database functions must be written that automatically update the shapefiles when triggered by any editing of the underlying data.

The storage of spatial metadata requires a specific data model that may not be part of the existing or commercial database product in use. The references included in Appendix A should be helpful in designing the extensions to the geotechnical database to accommodate the spatial metadata.

DIGITAL DOCUMENT STORAGE

At some point in the planning or implementation stage of a database project, someone will express keen interest in having various documents and digital images available as an integral part of the database. In most cases, these are valid items to consider as part of the database because hardcopy documents (maps, cross-sections, graphs, charts, and reports) are an important part of the geoscientist's interpretive process. However, incorporating digital images into a database brings into play a multitude of technical and logistical problems that many users have never been forced to consider. In this section, some of those issues are considered and possible solutions presented.

Objectives of digital document storage

Every user has his or her concept of how digital documents should be used. Digital documents can be used to archive irreplaceable or easily damaged paper copies, as compact replacements for bulky paper documents, or to provide quick access to a central document repository from a remote location. In addition to the overall objectives of the data management project (see *Defining Project Objectives* in *Planning Database Projects* chapter), certain logistical considerations must be addressed at the beginning of a project that includes digital document storage.

Sometimes it would appear that a document-imaging project is conducted because someone (usually at a high management level) thought it would be neat to have all corporate documents in a digital format. This line of thinking seems to equate the use of technology with being high-tech. It is better to fully understand the "why" of this issue very clearly at the outset because document imaging projects usually represent a significant investment in time, money, and personnel. Some of the basic questions to be addressed could include the following:

- What is the purpose of replacing hardcopy documents with digital versions? Is the primary reason for more rapid access by multiple users, or is it to preserve and archive aging or irreplaceable paper documents?
- What will be done with the original documents once the scanning has been completed? If the originals will be destroyed, the validation and quality control of the digitizing process must be absolute.
- Will simple images be sufficient, or will the scanned images need to be converted to digital data? Conversion of scanned images requires special optical character recognition (OCR) software and may require different scanning parameters. The converted data must also be quality checked and validated with a much higher degree of scrutiny than simple imaging.
- How much real cost savings will be generated by a scanning project? Factors such as time, manpower requirements, digital storage costs, document management software, maintenance, and support must be considered when determining the actual cost savings from a scanning project.

The fundamental workflow used in the conversion of hardcopy documents to a digital format can be fairly complicated, but the basic steps can be as shown schematically in Figure 7–9.

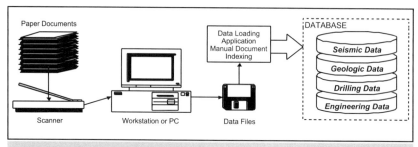

Fig. 7–9 General workflow for document digitization. The actual workflow will be determined by the hardware and software used, as well as special document requirements (oversize maps, well logs, etc.)

Paper documents can represent virtually any hardcopy information, including maps, cross-sections, reports, log headers (log curve data are normally handled differently), and graphs. Individual paper documents are scanned under control of a data scanning and loading application (and hopefully under the supervision of a qualified, experienced operator), indexed, and stored in the database. Although the process seems simple and straightforward, the most critical step in the process is indexing and storing the scanned images. Without accurate and appropriate indexing, the process is a waste of time and money.

One must also ask who will convert the paper documents to digital format. While this seems like an obvious question, sometimes it is the most elusive. It is neither practical nor advisable for a secretary to embark on a project to scan 300,000 paper documents using a desktop PC and flatbed scanner. If the project is of any size or consequence, an outside consulting firm or contractor may be the best approach. Most qualified imaging companies can provide the staffing, hardware, software, and experience to provide a total solution that may not be available in-house. This does not imply that the contractor will have any idea what data are being scanned, as this falls into the realm of metadata (see below).

The following are some suggested guidelines to consider when evaluating a scanning contractor:

- What experience does the contractor have with the specific types of data that will be scanned? A contractor with limited experience may create more problems during the scanning process than the project is intended to solve.
- Does the contractor have sufficient resources (time, people, and hardware) to complete the project within the required timeframe? While adding hardware for a scanning project is relatively straightforward, locating experienced personnel and providing specific project training is a time-consuming, difficult process.
- How reliable is the contractor in terms of data quality, timely performance, and responsiveness to the client's needs? Talking to the users from another company who have used the contractor's scanning services will provide insight.

The next step will be to determine who will provide the metadata (indexing) for the documents. Scanning, indexing, and storing digitized documents assume that someone ultimately will want to retrieve the information. Unless a knowledgeable user applies standardized criteria to index the document, there is very little chance that the information can ever be retrieved. Building a metadata index for a scanning project will involve two key steps:

- *Defining the keywords* that will be used to identify the individual document. In many cases, these keywords will exist in some other database, but in most cases the user community will need to assist in defining the indexing keywords.
- *Creating a hierarchical indexing system*, with standardized lists of keywords (indexing values) for each level of the hierarchy. For example, to provide the minimum indexing for a geological map, one might need to index the map by geographic region, area, field, horizon, and map type. Additional indexing might include scale, author, date, and company. In some cases, 25-30 attributes are needed to provide an adequate index for a single document.

Be sure to determine how much physical storage space will be required. Paper documents are tangible, measurable, physical objects, and the storage requirements are generally well defined (i.e., there are always more documents to store than space available to store them in). How much space will be required to store a scanned image of the document, however, is a more elusive quantity. Space requirements will be determined by the scanning resolution, physical size, and color depth of the original documents. Some guidelines to consider are as follows:

- Do not scan pure text documents at high resolution (>200 dpi). The scanning resolution should be determined by printing a copy of the scanned image and comparing with the original. Most text-only documents require only a low-resolution scan.
- Avoid grayscale scanning and color scanning for pure black-and-white originals. The exceptions to this are very poor-quality originals where a grayscale image may be more legible than a pure black-and-white image.
- Avoid color scanning at extremely high resolution (there are various levels of color imaging, and the lowest resolution that produces an acceptable print should be used). Color scanned images are among the largest files and can significantly impact the overall storage requirements for the project.

Determining how much time will be needed to complete the project may be part of an estimate from a scanning contractor, or a rough estimate can be made based on the volume of documents that need to be scanned.

Be sure to ask yourself what application will be used to store and retrieve the documents. The choice of document management software requires consideration of several factors:

- Are the digital documents stored in the application in a proprietary format? If so, what are the capabilities of the application for exporting the documents in an industry-standard format? These questions are very relevant in the event the imaged documents must be migrated from one application to another. Being tied to a single-vendor solution is very dangerous without either a conventional internal storage format or the ability to migrate the documents easily. Should the vendor no longer provide support for the application, the entire document database could be compromised.
- Is the document management software compatible with the data management system being used to manage other geotechnical data? If so, is it possible to provide a direct link between the two systems?
- Does the software vendor provide ongoing support and maintenance? If so, talk to other clients who currently use the product to determine if the support is adequate.

Case history: document scanning projects. Anyone who has worked in the petroleum industry for any length of time has experienced at least one attempt at achieving a paperless workplace. Some of these efforts have met with varying measures of success, but many were doomed to failure long before they were initiated. This example presents two similar case histories that illustrate the range of problems that can be encountered in any document management effort.

The first example involves a foreign office of an international company that wanted to achieve a totally paperless office. Meetings were held at very high levels between the management (mainly IT staff and programmers) and various software vendors and consultants. Using the vendor-supplied data, an action plan was developed and sent to appropriate department managers for implementation. Recognizing the unrealistic nature of the plan, extensive user interviews were conducted to determine document types, data volumes, and document utilization. Once these realistic and rigorous estimates of time, cost, personnel, and hardware/software resources were compiled, it became clear that the original plans had underestimated the scope of the project by an order of magnitude. In this case, the higher-level managers trusted the department managers' estimates, and the project was revised and scaled down to a more realistic and achievable level. This isn't always the case, however.

Another example of the pitfalls of document management projects involved a national oil company that had already embarked on an extensive document-scanning project. In fact, over 500,000 images had already been scanned before users were allowed to access the digital image files. At that point, it became clear that metadata had not been collected (i.e., no keyword or other types of indexing) for the scanned images. Documents had to be located in a series of handwritten ledgers to find the filename of the scanned image. Again, management recognized the problem, and a team of users worked with the contractor to develop a more comprehensive workflow that would collect appropriate metadata at the time of scanning.

Both of these examples illustrate the need for proper planning. The planning must include realistic estimates of time, cost, and resource impact and should involve extensive interviews with the end users and current data custodians before starting the project. If a project of this type is already under way, it should be reviewed quickly for any potential problems such as these, with corrective action taken immediately if necessary.

8

Data Reformatting

Populating, or loading data into a database system often requires copying (migrating) data from various disparate sources or legacy databases. In these cases, it is usually necessary to perform some sort of data reformatting, modification, or transforming to ensure that all the data in the new, combined database system are internally consistent. This chapter introduces the basic concepts of data reformatting, presents selected case histories, and outlines suggested potential solutions for the most common data reformatting problems.

GOALS AND OBJECTIVES OF DATA REFORMATTING

Even with the integration and standardization of modern data management and intepretive applications, there will inevitably be cases where data need to be reformatted for another application. In some instances, the reformatting process is simple and automatic, but complex reformatting problems require careful planning and standardization.

Standardization goals

One of the purposes of reformatting data is to standardize the database contents. Smooth, efficient, accurate database operation requires that all contents (data) conform to specified standards for format, style, and content.

Data reformatting is commonly necessary to convert a data element from one format to another. Consistent field formats are necessary to make data accessible and compatible throughout the database. For example, consider a dataset being loaded from a legacy system to an existing database. In the legacy dataset, dates have been stored as text fields, but in the existing database they need to be stored in a date/time format. The legacy data therefore need to be converted from text to dates so that all values in that date field are in the correct format. Attempting to load text strings into a date or numeric field will almost always cause problems with import operations or the resulting data.

In some instances, data may conform to the data format required by the destination database but the content of the individual data fields is inconsistent with the rest of the database. For instance, the source database may contain a text field that includes characters that are not permitted in the destination database. In this case these characters would need to be removed from the source data to preserve the consistency of the destination database, even though the two data types are the same.

Application integration objectives

Another important objective of data reformatting is to better integrate the database contents with the application software. Because the specific format requirements of interpretive applications can vary between products, it may be necessary to provide some data reformatting while exporting or transfering data to the application. Maintaining various formats within the database is poor data management practice, and even with this approach special cases will require export data reformatting.

Specific application format requirements. When an interpretive application needs to access the data stored in the master database, several preliminary steps should be taken. Most of these can be accomplished during the planning stages of the database design and implementation; but for new applications, the same steps will apply.

The application software should be reviewed carefully to determine the specific import formats required for each data type used by the application. In many cases, the program documentation will provide this information, but it may be necessary to work with the software vendor for specific format details.

If a specific import format (such as dBase, Access, Excel, or LAS) is available in the application software, a corresponding standard data export file can generally be created by a built-in functions in the database system. If a custom format is required, it may be necessary to create a custom export file format from the database. In most cases, of course, this same template can be reused anytime that format is required.

If no existing data import format is available in the application, then a custom export format needs to be created in the database system. Next, a custom import template needs to be created in the application. Once a working template has been created and tested, it is helpful to store the template in a shared directory on the system so that all users can have the same template to import the same data. Publishing and maintaining a catalog of these templates for all users to access can help minimize duplication of effort and possible import errors.

Obviously, these procedures need to be done for each data type to be imported by an application.

TYPES OF REFORMATTING PROBLEMS

The following discussion covers the majority of reformatting problem types that will be encountered with geotechnical data. The solutions and suggestions included here should also be helpful in other situations not covered here.

Simple data manipulation

Most data reformatting problems are fairly simple, consisting of a minor change to the format or data type of a few data elements. These types of problems are easily solved using the built-in functions available in most contemporary DBMS applications. In special cases, it may be necessary to develop custom user-defined functions (which may use several built-in functions) to accomplish the job. This may require the assistance of in-house or consulting programming support.

Data format conversions are probably the most common type of reformatting problem. Several examples of this type of problem are discussed in more detail below. These conversions are generally done on individual data elements on a record-by-record basis in the database or in the flat file used for data transfer. More complex data reformatting problems may include multiple data element and file style or layout differences, making the problem much more difficult. In general, the most complex reformatting problem should be solved first; then individual data element format changes should be dealt with.

Complex reformatting problems

More complex problems are encountered when dealing with files exported from legacy systems where there is no possibility of modifying the export format. Complex reformatting problems range from simple removal of a short header section at the beginning of a text file to extensive reformatting of files that were created as reports with page breaks, header and footer information, mixed columnar output, and line wrapping problems.

Extracting header information. A typical complex reformatting problem involves removing one or more lines of header information at the beginning of a text file that is being imported to the database. While removing header lines is relatively easy using the import wizards of tools such as Excel, file size limitations sometimes make this solution impractical for large data files. In these situations it may be necessary to enlist the help of a programmer to write a simple application or script that removes the header lines, at which point the remainder of the file can be imported using more conventional approaches. Following are some factors to consider when working with a developer.

- Will the header information vary in length? If so, it may be necessary to incorporate a file-viewing option that lets the data manager, data loading technician, or DBA view the contents of the file, determine the number of lines of header information, and input to the program the number of lines to be removed.
- Is there important information in the header that must be retained? In this case, a more complicated application may be needed to extract the

needed data from the header portion and store it either with the actual file data or in another file for subsequent import to the database.

Removal of report formatting. This problem is among the most complex reformatting challenges. If the file is sufficiently small, it may be possible to import the file using Excel or an equivalent spreadsheet application.

During the import process, the file format should be set up to reflect the actual data portion of the file, which hopefully is in a consistent columnar format. Once the file has been imported to the spreadsheet, a series of record-sorting operations collect the blank records (blank lines in the input file) and extraneous records (header, footer, and other format lines), which can then be removed. For the most complex situations or for files that are too big to fit within the constraints of a spreadsheet, programming assistance will be necessary.

Data conversion with reformatting

During the reformatting process, it is possible to include data conversions and other data element changes at the same time. This is only advisable when the input files are consistent and the actual reformatting is relatively simple.

When multiple conversion and reformatting operations are necessary, a good strategy involves using a multistage approach. For example, one step in the process might remove unnecessary header lines. The next step might perform simple data format changes or any other necessary changes to make the file compatible with the import procedures and internal structure of the destination database.

Case history: log curve naming example. Migrating data from legacy data management systems can be one of the most challenging problems in this field. In most cases, considerable effort and expense have been put into acquiring and loading the data, and many times the legacy database is the only source of the information. With the massive downsizing, merger, and reorganization cycles over the past 20 years, it is also unlikely that there is anyone available to consult on the history of the legacy database and its content.

This example involves a domestic company with a historical log database being migrated to a more modern log DBMS. The only export capability provided by the legacy system is the ability to create a simple text file listing of user-selected data fields.

Normally, this would have provided the necessary information to migrate the data, but there were two major problems. First, the data records in the legacy system contained a list of all logs for a given well in a single record, while the new system required that each log type be listed on a separate record. Second, the available log types were listed as a concatenated text string in no predefined order and with no standardization to the log mnemonics. An example of curves available for a typical well might look like:

FDC/CNL/GR/LLS/LLD/CL/BHC

The solution required intermediate steps to convert the data from well-centric data to log-centric data, in addition to extracting the individual log types and standardizing them.

A custom (but disposable) solution was developed that would provide a multistage data conversion. The first step was to export the legacy database contents to text files using the available report-generation capability. Next, the individual data elements were loaded to the correct fields in an intermediate database table. Then, the individual log type names were extracted from the concatenated text string and a new record was created for each log type that was indexed to that well. At that point, another utility was used to scan for all occurrences of similarly named log types and to impose industry-standard log mnemonics. Finally, the properly formatted and standardized records were loaded to the new database.

Although considerable time was invested in developing this solution, its life expectancy was very short. Since the legacy database was no longer used (and not supported because the original vendor was no longer in business), the loading and conversion tools were not needed once the data had been migrated. This is another instance where proper database planning and standardization would have made the data migration process much easier and more rapid.

DATA FORMATTING STRATEGIES

The data formatting strategies and suggestions included in this section should provide a basis for solving most of the common problems associated with geotechnical data reformatting.

When and when not to reformat

There is a fine balance between the desire to maintain the simplicity and integrity of a database and the ability to provide data in various formats needed by user application software. As a general principle, only one format for any given data element in the database should be maintained. The format for each data element should be a function of the data model as well as the specific objectives of the database system. Converting data elements to other formats should be in response to the type of data being imported or the requirements of the application software receiving the data.

Reformatting on input. In most cases, data being loaded to the database that do not conform to the internal formats of the database structure need to be reformatted. The reformatting operation can be done in one of two ways.

If the data to be loaded to the primary database are being transferred from a legacy database system or from another system with a different data schema or structure, every effort should be made to export the data in a format compatible with the primary database before they are loaded. In some cases this is impossible because the data will be delivered in a predefined format that the DBM and DBA have no control over. In these cases, the reformatting will need to be handled during the loading or import process.

If the data to be loaded are already in a predefined format and there is no possibility of recreating the file in a different format, it will be necessary to make the necessary format changes during the import process. This can be done through a series of user-defined format conversion functions or through a multistep loading procedure that moves the data into a temporary table, makes the format conversions, and then loads it into the final destination table.

Reformatting on export. If the data are to be exported to an application program that requires the data to be in a modified format, the reformatting should be done during the export process. As in the import process, there are two basic options available to accomplish this.

The ideal method of exporting data to an application program is to create a specific export file format that generates a data file that can be loaded directly by the application program. This can be done using the report-writing tools in the DBMS combined with user-defined functions and macros that make the conversions on the fly. Where possible, these export templates should be standardized by data type and receiving application, and documented in a catalog of templates for reuse when needed.

The second alternative to custom file export templates is to move the selected data to be transferred to a temporary table, make the necessary format changes (again, using a combination of reusable user-defined functions), and export to a file for import by the application program. This method may be preferred where the application software is capable of reading data directly from the database tables, thus eliminating the steps of creating an export file and then importing it.

Regardless of the requirements of the application software using the data, the *original format* of the data should always be preserved. This is equivalent to preserving the original, observed, or measured data in the database as it represents the base data before any changes, modifications, or reformatting has been done.

Sorting vs. indexing data

One of the first operations that most people want to perform on any given dataset is to sort or reorder the information. This can be accomplished two different ways, and it is important to understand the fundamental concepts and problems associated with both approaches.

When dealing with a spreadsheet or table of data, most applications allow the user to *sort the data* based on the data contained in one or more columns of the table. While this produces the desired result (in most cases), the rows in the table are permanently reordered. Sorting the table several times, using different criteria can result in a row order that cannot be returned to the original order. This original record order may be important, as discussed in a case history below.

A better alternative to sorting is by *indexing* the table on a key field or column. In a relational database, a table needs to have at least one primary key (sort field) but can have multiple secondary or foreign keys (links to other tables). A primary key field can have no duplicate values in the table in which it appears. The important distinction between sorting and indexing is that the table can be reordered by an index field but the actual original record order is not destroyed.

The following example illustrates the differences between sorting and indexing and points out some of the problems associated with both approaches. Consider a data table of formation tops that contains between 8,000 and 10,000 records. A user adds an additional 1,000 records that contain inaccurate revisions to existing rows in the table. After adding the rows, the user sorts the data by formation name and well name. At that point, the errors are discovered; but because the table had been sorted, the 1,000 errors are now scattered at random throughout the table. If the table had been indexed instead of sorted, the errors could have been removed by eliminating the last 1,000 rows in the table.

EXAMPLES OF REFORMATTING SOLUTIONS

The examples discussed in this section illustrate solutions for common reformatting problems associated with geotechnical data. These should serve as guidelines for more complex or more unique problems.

Converting text to dates. Some database systems (Access, for example) automatically convert text strings that look like dates into a date-format field. There is one cautionary note here, however, on date formats. The date format most commonly used in North American is mm-dd-yyyy or mm/dd/yyyy (although there are those who still insist on using only two digits for the year). However, in Europe and much of Asia, date formats are more variable, and can be written dd-mm-yyyy, yyyy-mm-dd, or even yyyy-dd-mm. For example, the date May 12, 2001, can be expressed as 05-12-2001 or 12-05-2001, depending on local convention. Obviously, it is important to know what the intended date format is for the data being used, especially before letting an automatic reformatting wizard load the data for you.

Assuming that one does not have access to an automatic date format conversion utility or that the dates are in an unusual format, the dates may need to be converted manually. One way to accomplish this is to use a series of built-in database functions to manipulate the text string to convert it to the correct format. Using a substring text function (similar to the InStr function in Access) allows the original string to be split up into its component parts, then reassembled into the desired format.

Converting numeric data to special text formats. There are certain circumstances where numeric data need to be formatted into a more human readable format, especially for formal reports and data listings. One good example of this problem is the presentation of coordinate data expressed as latitude and longitude. While the database should contain the coordinate values in separate numeric fields for degrees, minutes, and seconds, a more readable format shows the data as a single text string with the appropriate symbol notation for degrees, minutes, and seconds.

Case conversion. Another common format conversion is the simple reformatting of text strings from one case (capitals) to another (lowercase). Commonly, short text fields (especially those used as primary key or foreign key fields) are expressed in all capital letters to ensure consistency during search and query operations. However, when these fields are presented in a formal report, the use of all caps may not be the best look. The use of simple built-in case-conversion functions—like LCASE(), UCASE(), and PROPER() functions in Access®—can make the conversion to the desired presentation format without affecting the format of the data stored in the database.

Converting text to numeric data. In some cases, it may be necessary to extract portions of a text string and store the individual portions in numeric-format fields in the database. To illustrate using the above example of formatted coordinate data, consider the situation where coordinate data are delivered in human-readable format and need to be split into individual numeric fields (degrees, minutes, and seconds) and stored in the database. There are various methods to extract the individual parts of the text string using built-in DBMS functions.

Once the data have been reformatted to a stucture compatible with the data management system, they can be imported or otherwise loaded to the database. The next chapter discusses data loading and input, and assumes that any data formatting problems have already been resolved.

Data Loading and Input

After any reformatting problems have been identified and resolved, loading legacy data to a new database can be accomplished in many ways. Moving data within the same suite of applications is generally straightforward but can become more complicated with each change of operating system, DBMS, or other variation. Although direct input from the user is usually easier, data validation and overall quality control must be considered. This chapter presents some of the most common methods of loading data to a new database as well as the basic concepts of direct user input methods.

SAME-SYSTEM DATA TRANSFER

By far the easiest method of moving data into a database is by transferring it from another part of the same database or from a directly compatible interpretive application. Even after the database is populated, there will continue to be a need to transfer data into the database either through direct input or from other applications.

Commercial examples

Most commercial geoscience interpretive applications provide some element of underlying data management support. The two most popular and widespread application suites from Schlumberger GeoQuest and Halliburton Landmark are discussed here. In general, other application packages generally provide some sort of data transfer capability to one or both of these systems.

GeoQuest Finder®. Schlumberger offers a wide range of interpretive and visualization applications intended to share the same underlying data. The underlying data management product, Finder®, was originally designed as a map-based database interface that lets the user search, visualize, and transfer selected data to a project database in one or more interpretive applications. These transfers are accomplished using the GeoShare® data transfer protocol.

Moving data from the Finder® data model to and from the related GeoQuest® interpretation and visualization applications is fundamentally a file transfer protocol rather than a direct movement of data from the database to the project. The GeoShare® process involves first creating a selected set of data for transfer to the interpretive application (either through the Finder® interface or several other GeoQuest® data browsers). Next, the selected data are moved into an RP-66 format data file (basically a standardized text-format file) that is transferred to the destination application. Finally, the application imports the data from the RP-66 file into the project database where it can be used for interpretation. Transfers back to the main database are done using the reverse of this process.

There are some drawbacks inherent in this process:

- Not all the data from the source database are transferred to the destination application. Any additional data items that need to be moved to the destination application must be transferred in a separate operation using some means other than GeoShare®.
- Once the data have been moved to the destination application, they remain out of sync with the master database in Finder® until they have been transferred back to the master database. This can create problems in that modifications may be made to the master database that are not reflected in the project data.
- GeoShare® data transfers only support a limited suite of applications (mainly GeoQuest® products). To transfer data to an unsupported application requires the development of specific loaders for that application using the optional GeoShare® Developer's Kit.

Landmark OpenWorks®. Halliburton's Landmark suite of interpretive applications takes a somewhat different approach to the transfer of data. The underlying OpenWorks® database functions as the master data repository for all interpretive and visualization tools in the product line. The OpenWorks® data manager has a map-based interface as well as the tools necessary to create and manage project datasets.

Moving data between the OpenWorks® database and the interpretive application is straightforward. Once the desired subset of data has been selected from the master OpenWorks® database, it is moved to the application using a pointer-dispatcher approach. The destination application is set to listen, and the dispatcher is activated in OpenWorks®, enabling the data transfer.

Landmark provides support for most major interpretive applications using a series of half-links or half-hooks. These transfer protocol applications represent one-half of the process. The origin database (OpenWorks®) contains one-half of the link, and the receiving application contains the other half. This procedure moves data directly from the database to the application (or back) without resorting to an intermediate step of creating a transfer file.

Like GeoQuest®, the Landmark methodology has its drawbacks as well:

- Once data have been loaded to the project database, individual data fields are not always directly updated when changes are made in the master data in OpenWorks®. However, if revisions are made to the master data records, another point-dispatch operation will update the project data as well.
- OpenWorks® may not be ideally suited for large, shared databases for a variety of reasons, performance being chief among them. To optimize the performance of OpenWorks® it is generally necessary to split the master data into regional or asset-oriented subsets, which can then be shared by a smaller number of users. This results in a number of master databases, complicating the overall maintenance and enforcement of standards.

Many large DBMS vendors are now supporting the OpenSpirit development environment. OpenSpirit is an industry-supported consortium that promotes an open development environment for a database-to-application interface. At this time, connections to the underlying databases are read-only (i.e. changes in project data are not replicated directly in the master database). In the future, database connectivity issues may be significantly reduced through these types of initiatives.

INTER-DATABASE DATA TRANSFER

If it becomes necessary to transfer data from one database management system to another, the best approach is to use a variety of tools available to provide this link automatically. Many of these tools are platform restricted; but even in situations where more than one computing platform is in use, solutions are available.

ODBC and SQLNet® links to tables

Tools such as open database connectivity (ODBC) drivers and SQLNet® let users of one database system connect to and use data from other databases. The most common of these uses is the connection of PC-based databases, such as Access® with server-based master database systems such as Oracle. Within certain limits, these connectivity tools make direct and seamless links between the two databases.

ODBC. The Open Database Connectivity (ODBC) protocol developed by Microsoft allows PC-based computing systems to link to a variety of other database systems. Like most things Microsoft, this is accomplished using a specific ODBC driver for the remote database. To access another database using ODBC, the user establishes a link to the remote database from the PC database by selecting the appropriate ODBC driver for the source database. While Microsoft and other vendors provide links for a wide variety of database products, this is not a universal solution. If connection to the database from client PCs will be an integral part of the database system, an ODBC driver for that database is essential. If this is not available out of the box from Microsoft, the DBMS vendor should provide this functionality if possible.

SQLNet. Applications like SQLNet allow PC and UNIX platform users to access Oracle databases directly in much the same way that ODBC drivers are used. The major difference is that these applications provide a front end for extracting, viewing, and editing data in the host database without actually making a direct link to the database (as in the case of ODBC).

Drawbacks to direct links. While the convenience and simplicity of ODBC and similar access methods are attractive, there are certain limitations and problems inherent with this approach.

- ODBC drivers are only suitable for accessing other databases from a client PC running a Microsoft operating system. Some older operating systems may not be compatible with current ODBC drivers.

- Extracting data using an ODBC link is simple, but a major problem develops when the user needs to move data back into the master database. Unless the proper authorizations and data validation tools are provided, data movement is a one-way street. Also, giving a user direct access to the host database may circumvent carefully constructed validity checks and input filters. The obvious result will be corrupted or, at best, suspect data.

Exports to DBMS format

If it is impossible to link directly between two databases, another approach is to use the native export functions within the DBMS itself. Most relational database systems have built-in functions to create export files that are native to that particular database system. Normally, these functions are used by the DBA to create backup or archive files of the entire database or limited subsets. However, this approach can be used if moving large amounts of data between two installations of the same DMBS.

If you consider this approach for moving data between databases, work closely with the DBA responsible for the database. The appropriate version number and format of the export file must be compatible with the receiving database.

Export to flat file (text)

As a last resort, data can be moved between databases using flat files. Although this is a slow and tedious method, it provides complete control over the export and import process and generally works.

When using this method, there are two very important things to bear in mind:
- It is always advisable to export the entire contents of every table being transferred into a separate flat file. Even if not all of the data are being loaded to the receiving database, the export files are always an exact and complete copy of the original in case additional data need to be extracted and loaded.
- The table structure should also be exported to a flat file to ensure the destination database table has the exact same structure, field names, key and index identifications, and data formats.

The format of the export file should be as simple as possible and should adhere to the following general guidelines.

Export files should be standard ASCII text and normally should be a delimited format. The choice of delimiters is optional, and a comma, tab, spaces (blanks), or other delimiting character can be used as the field delimiter (separator between fields). The import capabilities of the destination database may determine which of these delimiters is appropriate or if a columnar format would be better. Also, if the data fields being exported contain spaces, commas, tabs, or other characters normally reserved as delimiters, using a fixed-format export (where each data field appears in a predefined position in the file) may be a better choice.

Each record in the source database should be written to a *single line* in the export text file. This can present problems with extremely large tables, which may require that the table be broken into smaller parts before exporting.

Fields in the source data table containing *text strings* should be quoted so that any spaces, commas, or other special characters in the text string are not confused with a field delimiter during the import process.

Columnar files are also referred to as fixed-format files. If a columnar format file is more easily imported by the destination database, or if the data contain special characters, this style of export file is more suitable. However, columnar format export files must conform to the same general guidelines as delimited files. Leave enough space between adjacent columns so that there is no confusion where one field ends and the next begins. An added benefit of using a columnar format is that the file is easier to read with a text editor, should that be necessary.

The *column (field) headings* should only be placed on the first line of the file unless the importing program can distinguish the first line as column/field headings. Otherwise, the field headings (the actual field names in the source table) should be written to a separate file—if possible, with the field length and data format for each field. In a columnar format file, numerical data should be written with a constant number of decimal places and right justified. Text fields should be left justified in most cases.

Limitations of flat-file transfers. While flat-file data transfer is fairly simple and easy to use, it can be time consuming and requires a great deal of effort and user intervention. Other potential limitations are as follows.

- Source tables with a large number of fields may be difficult to transfer directly using this methodology. Since the normal length of a single record (line) in an ASCII format text file is 256 characters, the

total length of the export record in the flat file cannot exceed this total. If this is the case, it may be necessary to break the table up into more than one export file and then recombine the parts during the import process.
- Flat-file data transfer does not move the entire source database schema and does not maintain any of the information regarding table relationships, views, indexes, or other information managed by the DBMS.

LOADING FROM TEXT FILES

Virtually every data management implementation will require loading data from various forms of text files. In most cases, text files contain data that cannot be loaded to the system directly or data from legacy systems that may not be available in any other format.

Major import considerations

While many methods and applications can be used to edit text files, there are certain common problems, considerations, and suggestions that apply to all text-import issues.

Physical file size. Many text-format data files are very large, and some text editor applications cannot accommodate the physical size of the files. Even if the editor can open a file, navigating within the file and the time involved to import the file (especially with a slower computer) make using large files impractical. In some instances, it is better to break the data into smaller, more manageable files.

It is also advantageous to use a small but representative data file when testing data import routines or conversion utilities. The pilot datasets that the users provide during the planning phase may be ideal for this purpose.

File format considerations. As discussed in the preceding section, text-based data files can be delimited with a variety of characters or columnar format. Unless the text files are created following the above guidelines, it may be very difficult to determine the actual file format, field names, data formats, etc.

Column formats. Columnar format text files are much easier to read than delimited format files, but they usually limit the number of fields that can be placed on a single line (record) because of the white space needed to create the column format. If importing a columnar format data file, it is always best to view the data in a text editor before running the import process to ensure there are no extraneous characters, extra lines, or misaligned columns. If the columns were not properly justified, it may be necessary to adjust the import column widths to account for column alignment problems.

Line wrapping problems. The most common problem with text format data files is the introduction of inadvertent line wrapping, where records are truncated and wrapped to the next line in the file. This can be prevented by adjusting the export file to make sure the text does not wrap. In the worst case, it may be necessary to open the export file in a text editor, search for, and edit the wrapped lines so each record is contained on a single line in the file.

Generalized import procedures and solutions

In many large data management projects, importing data from large legacy data files is a very common operation. To minimize the number of import errors and reduce the time involved in creating import (migration) tools, several general steps should be considered.

Remove line wrapping. While ASCII text files all appear to be the same, various subtypes have characteristics that can present problems when importing the text to a data management system. The most common of these is the effect of line wrapping on the individual lines of text. In most cases, each line of text is assumed to represent a single record in the database table. As such, if the text is truncated or wrapped to a new line, there will be extraneous characters (i.e., the ASCII code for a new line) at the end of each line. Furthermore, the next line in the file will not be a new record, but will be the wrapped portion of the preceding record.

This problem can be avoided in several ways. The simplest solution is to make sure the application used to import the file does not wrap or truncate the text. If it is necessary to work with a text file where wrapping has already been embedded, do a global search and replace, where the "search-for" text is the new line code and the replacement text is a null string (Fig. 9–1).

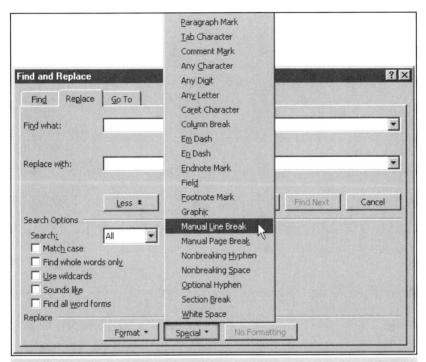

Fig. 9–1 Using the MS-Word® text editor to search and replace manual line breaks in a text file. Most word processing software and spreadsheet applications have similar search-replace dialogs.

Removing page breaks. Many exported text files were generated by a report-writing routine intended for printing of hardcopy. In these cases, the text file often contains hard page breaks (coded printer instructions that force the printer to eject a page and continue printing on the next page). Most text editors can search for the embedded code that forces page breaks. Searching for the hard-coded page breaks and replacing the code with a null string will usually correct this problem.

Removing extraneous/blank lines. In addition to page break instructions, many text files also include headings, page numbers, and other nondata items. It is very important to remove this unwanted information before attempting to import the data. This can be done in a spreadsheet or by using custom scripts or file-operation macros.

Pilot data files. Once the input data file has been cleaned and the import application or template created, it is helpful to create a small subset of the complete file. This allows rapid testing and debugging of the import procedure, which is especially helpful if working with a slow processor or multiple, large files. It also protects the master data from unexpected problems. The only consideration in preparing a pilot data file is that the subset file provides an accurate representation of the data found in the total file.

Data type changes. A very common problem with importing legacy data files is that data types must be made compatible with the new database system. Often the import routine is limited to a simple partitioning of the lines of text into smaller chunks of text without actually converting the data into the new format. One way to solve this problem is to use import wizards that allow the user to specify not only the beginning and end of a data field but also the data type to which that the field should be converted.

Data conversions. Once legacy data have been imported to the new data management system, there are always situations where data need to be converted from one format to another. When faced with this problem, your easiest solution is to first create a new field in the data table that has the desired format for the converted data. Next, the database system tools and functions can be used to convert the data to the desired type and store it in the new field.

The conversion should be checked at this point to make sure it is functioning properly, especially if the operation is complex or requires a user-defined conversion function. Once the user is satisfied that the conversion has been done correctly and completely, the original field can be deleted from the table and the new field renamed to the original field name. This is fundamentally the same procedure as changing field names.

Some database systems let you directly rename fields in a table. This operation should be tested on a copy of the original table first unless the user is certain the renamed field will not be corrupted during this operation. If there is any doubt that this operation can be done without errors, a three-step procedure should be followed.

1. Create a new field in the same data table with the same format and other parameters as the field to that is to be renamed.

2. Copy the contents of the original field into the newly created field. The new field should be named whatever the new field name should be. Verify that the contents of both fields are identical.
3. Delete the original field from the table.

The following series of screenshots illustrate this procedure with a simple table containing well name, formation top name, and depth.

Top_Name	Well_Name	Depth
▶ Cretaceous_A	Well_011	5640.2
Cretaceous_B	Well_011	5791.1
Permian_Top	Well_011	5800.2
Triassic_Base	Well_011	6010.7
Cretaceous_A	Well_015	5789.0
Cretaceous_B	Well_015	5840.5
Permian_Top	Well_015	5901.9
Triassic_Base	Well_015	6113.4
*		

Fig. 9–2 Initial table contents and field names before any revisions have been made.

Figure 9–2 shows the original table before any revisions have been made. In Figure 9–3, the original table has been revised to add new fields to the table structure that have the desired field names.

Top_Name	Well_Name	Depth	Formation_Name	MD_Feet
▶ Cretaceous_A	Well_011	5640.2		
Cretaceous_B	Well_011	5791.1		
Permian_Top	Well_011	5800.2		
Triassic_Base	Well_011	6010.7		
Cretaceous_A	Well_015	5789.0		
Cretaceous_B	Well_015	5840.5		
Permian_Top	Well_015	5901.9		
Triassic_Base	Well_015	6113.4		
*				

Fig. 9–3 New fields have been added to table with the desired final field names.

After the table structure has been revised, the contents of the original fields are moved (copied) to the new fields using direct SQL commands or an update query, as illustrated in Figure 9–4.

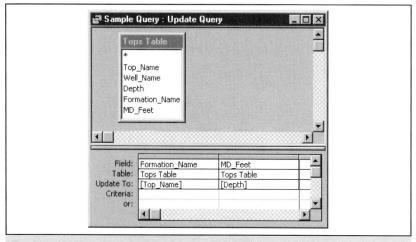

Fig. 9–4 Update query designed to move the original data into the new fields.

The results of the update operation should be verified to ensure the original data have been migrated correctly to the new fields, as shown in Figure 9–5.

Top_Name	Well_Name	Depth	Formation_Name	MD_Feet
Cretaceous_A	Well_011	5640.2	Cretaceous_A	5640
Cretaceous_B	Well_011	5791.1	Cretaceous_B	5791
Permian_Top	Well_011	5800.2	Permian_Top	5800
Triassic_Base	Well_011	6010.7	Triassic_Base	6011
Cretaceous_A	Well_015	5789.0	Cretaceous_A	5789
Cretaceous_B	Well_015	5840.5	Cretaceous_B	5840
Permian_Top	Well_015	5901.9	Permian_Top	5902
Triassic_Base	Well_015	6113.4	Triassic_Base	6113

Fig. 9–5 Results of update query operation with the original data migrated from the old field names to the new fields with the desired field names.

Finally, the old fields are removed from the table, and the field order is modified (if desired) to yield the final table structure with the new field names (see Fig. 9–6).

Well_Name	Formation_Name	MD_Feet
Well_011	Cretaceous_A	5640
Well_011	Cretaceous_B	5791
Well_011	Permian_Top	5800
Well_011	Triassic_Base	6011
Well_015	Cretaceous_A	5789
Well_015	Cretaceous_B	5840
Well_015	Permian_Top	5902
Well_015	Triassic_Base	6113

Fig. 9–6 Final table contents and field names (original data have been preserved, but the old fields have been removed.)

Case history: custom data loading solution. Here is a further example of disposable code. A large independent oil company was using commercial software for geocellular reservoir modeling and simulation. One of the data elements needed for reservoir characterization was information from core analyses and physical core descriptions. Core data are predominantly discontinuous, but the modeling software could only load continuous borehole data (i.e., wireline log data).

To solve this problem, a custom application was written that read the discontinuous core data and generated an LAS-format data file that inserted null values in places where no core information was available. The resulting data file was then easily imported to the application.

This utility was discussed with the commercial product vendor, but several years passed before the functionality was available in the product. Unfortunately, vendors often focus on developing new, highly visible functionality at the expense of solving more specialized existing limitations. While the commercial and competitive nature of the petroleum business cannot be ignored, software vendors should take better advantage of client-developed custom solutions.

After the data have been successfully loaded to the database, and any modifications or changes have been made to the table structures, it is necessary to ensure that the data have been normalized. The next chapter discusses the concept of data normailiztion and introduces methods to ensure that all the database contents have been standardized.

10
Data Normalization

Data normalization is a process by which the contents of the database (i.e., data) are made internally consistent. While much of the normalization can be done using data input validation (see next chapter, *Data Validation, Editing, and Quality Control*), this is not always possible when data are imported from external files, data tables, or legacy databases. However, the process is essential for successful data extraction, manipulation, reporting, and utilization.

This chapter defines the basic process of data normalization, explains why the process is so important, and provides examples of problems that can arise when normalization is done improperly. Several examples of normalization methods are demonstrated and compared.

DEFINITION AND IMPORTANCE

Database normalization refers to the internal structure and makeup of data tables and the links between them. Database normalization is a complex issue, but one that is generally addressed by the DBA. The basic goal of database normalization is to ensure that key data elements are maintained in a regular, organized manner with no repetition or unnecessary duplication of information from table to table within the database.

The concepts of *data normalization* are not as well defined as formal database normalization. The DBA ensures that tables are fully normalized and that the system is tuned; responsibilities rarely extend to the issue of data accuracy. A successful and useful database is one where the data are internally consistent and accurate. Internal inconsistencies and errors generally bring into question the accuracy of the entire database. A single major error causes the users to lose confidence in the entire database.

Importance to Database Effectiveness

If the information contained in the database does not conform to standards of format and content, the database becomes compromised. Data not conforming to standards may not be found by search and retrieval operations, and errors may occur when loading data to fields if the input data are in the incorrect format.

If the database information has not been normalized, it is sometimes impossible to make accurate and/or complete retrievals from the system. For example, if a user is searching for all wells drilled by "Exxon Corporation," the retrieval will not find those wells drilled where the operator name was entered as "Exon," "Exxonn," or some other incorrect variation.

Inconsistent calculations. One of the problems with geoscience data is that geoscientists are sometimes the only ones who can recognize incorrect or questionable data. For example, if a data table contains porosity data, all values must be entered on the same relative scale (that is, 0.10 is not the same as 10.0 unless it is clear that the units of that field are percent and not decimal percent). Data tables can be defined to have quality control limits so that porosity values can be entered only if they fall between 0.00 and 1.00 (i.e., decimal percent). However, most geoscientists recognize that any value beyond 50-60% (or was that 0.50-0.60?) is unreasonable.

This problem becomes very clear when importing legacy data from multiple sources or multiple users. Assuming the individual source or user was internally consistent is simply not enough. One source might have entered porosity values as percent and another as decimal percent. The resultant database will have both ranges, and any computation based on those values will be incorrect. Sometimes these errors can be much more subtle than this example and as such are very difficult to find and correct.

Integration with other products. Incorrectly formatted or unnormalized data can make direct integration with interpretation applications very difficult, if not impossible. If the application requires certain data items to be available in a specified format in the database (or in an export file), the data cannot be used if they are not in this format.

METHODS OF NORMALIZATION

Data normalization requires that inconsistent data be identified or located within the database. These inconsistencies must then be corrected (edited), and strategies must be developed and implemented to prevent the problems from recurring in the future.

Identifying data inconsistencies

The most difficult part of normalizing the contents of a database is locating all records that do not conform to the designed standards. It is always a case of "If you don't know what you're looking for, how will you find it?" Since there are no direct database functions that can be set to "find all the errors and fix them" (much as we would like to have these tools), this is another justification for having a DBM with a working background in the geosciences. Knowing where potential errors could be found and knowing what "reasonable and expected" values to look for allow the DBM to search effectively for possible problems.

Most databases have built-in functions that allow simple searches of the data for the most obvious problems. For example, simply indexing a table on the operator name will put all similar names together in the same place. Visual inspection will identify many errors using these simple approaches.

SQL approaches. SQL data retrievals can take advantage of a number of powerful built-in tools and functions. A simple example (again using the case of multiple spellings for the same operator name) takes advantage of the LIKE operator available in most versions of SQL. To retrieve all the records in the WELL_MASTER table where the operator name (OPERATOR_NAME) is something like "Exxon" would take the form:

SELECT ALL FROM WELL_MASTER WHERE OPERATOR_NAME LIKE ("Ex") ;

This creates a list of all records that begin with "Ex," regardless of whatever else is contained in the name. To broaden the scope of this search and find all records where the operator name contains the letter "x," another function could be used:

SELECT ALL FROM WELL_MASTER WHERE InStr(OPERATOR_NAME,"x") > 0 ;

In this sample of SQL, the InStr() function searches the field OPERATOR_NAME for any occurrence of the letter "x" in the field. If the query finds an "x," the position of that letter in the field is returned by the function (if there is no "x," the function returns a zero, which is why the test for "greater than zero" is used). Of course, in this example any operator name with an "x" in it would be found, but it serves to illustrate the principle.

Search-and-replace strategies

Sometimes normalization can be as simple as a search-and-replace operation, common to most word processing software. While generally not as simple and straightforward as a word processor, the fundamental strategy is the same. Many database management systems have built-in tools that function very much like the same operation in a word processor. If these are not available, an alternative is to use internal (SQL) functions or a product developed specifically for that purpose.

The following discussion shows the basic steps taken in a general search-and-replace operation. This example uses SQL, but the basic steps are the same no matter which methodology is used.

1. Identify the target information that is incorrect or that needs to be otherwise changed. This can be done using a variety of SQL commands and/or statistical operations.
2. Specify the information that will replace the target data. In some cases it may be necessary to replace a subset of the data (e.g., replace only part of a text string).
3. Set any limitations on the replacement operations. This can be done using the SQL WHERE clause in the operation.
4. Issue the replace command.

When dealing with large amounts of data or where the search-and-replace operation is very complex, it is usually a good idea to make a dry run that only selects (retrieves) the data and shows what the result will look like. That way, if there is a problem with how the statement is written, it can be corrected before the actual search and replace is performed.

To illustrate, the following SQL command searches the field OPERATOR in table WELL_MASTER and replaces the operator name with Exxon for any record where the current value of that field is Exon. In the first SQL command, the effects of the replace operation are displayed by retrieving all records that will be replaced, but the replacement is not made:

```
SELECT WELL_MASTER.OPERATOR FROM WELL_MASTER
WHERE WELL_MASTER.OPERATOR = "Exon";
```

In the second SQL example, the actual replacement is made:

```
UPDATE OPERATOR SET WELL_MASTER.OPERATOR = "Exxon"
WHERE WELL_MASTER.OPERATOR = "Exon" ;
```

If the appropriate tools and methods are not available within the DBMS, it may be possible to adapt external solutions or purchase third-party developed solutions that will provide the desired functionality. Alternatively, if in-house programming assistance is available, it should be possible to develop the needed tools on a proprietary basis.

Automating the process

While the above examples illustrate the basic methodology that can be used to perform selective data normalization, manual operations of this type can be time consuming and tedious, and they can introduce unwanted errors into the database. When the same types of normalization are done repeatedly, it is best to develop some sort of automated procedure or program that will simplify the process, reduce time, and mitigate the risk of introducing errors.

Programmatic solutions. Even if all normalization work is done entirely within the data management system, some automation to this process can be accomplished through programmatic solutions. This normally requires the support of a developer—either in-house or contract. As with most parts of the overall data management process, careful planning will lead to a successful result if you include the following steps.

- Define exactly what the automation process needs to accomplish. Examples of the types of normalization processes needed should be provided in detail.
- Create a test dataset that accounts for all currently known problems in the database so the application can be tested on test data instead of the actual production data.
- Work closely with the programmer(s) during the development process to ensure the design specifications are being maintained in the implementation.
- Thoroughly test the prototype application on the test dataset before deploying the application for use with the production database. Carefully monitor the application initially to ensure that there no obvious "undocumented features" (bugs).

In some simple cases, it may be possible to handle most, if not all, normalization work with user-written macros, scripts, and user-defined database functions. A complete discussion of macros and user-defined functions is beyond the scope of this text, but most users are familiar with the Microsoft approach to macros and functions.

Most of the cases discussed in this chapter deal with data inconsistencies that have somehow been loaded to the database. The real goal, of course, is to prevent these problems from entering the database in the first place. Most, if not all, data problems can be eliminated during data entry and/or loading using properly designed data validation methods. The next chapter deals with data validation, editing, and quality control in detail.

11

Data Validation, Editing, and Quality Control

Even the best-designed, mature database occasionally requires some type of data validation. Although most data validation problems can be eliminated on the "front end" during data entry and loading, some spurious or erroneous data always manage to slip into the system somehow. This chapter presents some common-sense approaches to validating data within a database using the tools that are most commonly provided with DBMS software. These concepts are extended to the data entry validation concepts presented in the next chapter.

DEFINITION AND IMPORTANCE

The process of cleaning up the contents of a database falls under the general heading of data validation. Part of this process involves data editing; but in this instance it refers to editing that is intended to correct bad data, not changes or updates to the data. (That sort of editing is a continuous process under the control of the users.) No matter how much effort is put into planning and implementing a data management project, if the contents of the database are unreliable, the users will quickly abandon the database. Therefore, the overall success of a data management project is directly dependent on the quality control that goes into maintaining it.

Methods of Validation

Data content validation is normally conducted in two phases. The first phase is the initial data validation, which is normally done during or shortly after the initial population of the database. The second phase is the more difficult process of ongoing data validation. This ongoing phase is often referred to as data quality control and requires continual or periodic sweeps of the data to ensure that invalid data have not slipped past the data entry and editing validation checks. During these periodic reviews of data quality, parameters, business rules, and workflows for additional data entry/editing validation can be identified and implemented.

General validation methods

Certain aspects of data validation apply to virtually all data types. These are general concepts that can be used as a starting point when developing the methods and techniques for initial and ongoing data validation.

Most data validation should begin with a search for obvious bad data. For most geoscience data types, maximum and minimum reasonable values can be defined. The first step, then, is to search the database for any values that fall outside these reasonable and expected ranges. Although this approach won't identify data within the normal range that are incorrect, it will eliminate absurd data or data that are not technically feasible.

Illogical data points are the next target of the validation process. The DBM must have a basic familiarity with the type of data in use so that a logic-based set of validation rules can be established. For example, logic would indicate that a porosity value of 35% (while not unreasonable) in an otherwise low-porosity carbonate formation is not logical.

Validation of specific data types

Other fundamental validation rules can be developed for specific data types.

The most effective method of validating date format information is through the use of input forms (see next chapter). Using a pop-up calendar from which the user selects a date allows the system to store the selection in the appropriate internal format. An alternative method could allow input of the day, month, and year as separate fields, each with its own validation rules.

For virtually all depth-related borehole data, the simplest form of validation rule is to see whether any borehole depth data fall somewhere between the surface elevation and the total depth of that wellbore. This doesn't rule out incorrect depths—only those that cannot exist spatially outside of the range of that well.

Validation of numeric data is entirely dependent on the range of expected values for the specific data. The minimum and maximum (and default) values can be determined from specific user input, from statistics from existing data in the database, or from published sources of information (rock catalogs, logging chartbooks, etc.).

Formation tops problems

Stratigraphic data validation can present several unique problems, depending on the complexity of the data model. If a relatively simple data model is used, most of the data quality checks can be done fairly easily. Again, creating logic and data-dependent validation rules does most of the validation.

Tops below bases. If the fundamental depth check has been performed, checking formation intervals for consistency is the next step. If the top of a formation is below the corresponding formation base, there is obviously a problem. If the zone is overturned, the top and base should still be in proper descending depth order. However, an exception flag must be set, indicating that the zone is overturned (see "Stratigraphic Exception Codes," below).

Missing tops and/or bases. In a stratigraphic sequence, the top and base of any defined zone must be specified individually. A zone with only a top depth must be identified as a marker rather than a zone (which has thickness).

Measured and vertical depths. Unless the wellbore is perfectly vertical (which is highly unlikely), there will be a difference between the measured depth (MD) and the true vertical depth (TVD). If both MD and TVD are the same, several possible problems are indicated:

- The TVD has never been computed for the well.
- The directional survey data are missing for the well, so it is impossible to compute a TVD from the MD.
- The TVD has been computed incorrectly.

In addition, if the TVD is greater than the MD, it is possible that the calculations were done incorrectly, the two values are reversed, or there is a fundamental problem with the directional survey data or reference elevation.

Stratigraphic exception codes. Stratigraphic exception codes (SEC) should be used wherever the stratigraphic sequence is abnormal. Examples of situations where stratigraphic exception codes would need to be used are as follows.

If the SEC indicates that the section is removed by *faulting*, the overall zone thickness should be less than the statistical mean for the reservoir in that area. Conversely, if the section is indicated as repeated, it should be thicker than expected. The reverse is also true: a thinner- or thicker-than-expected section should have some sort of fault exception code.

Subtler than the case of a faulted section is an erosional unconformity. The zone thickness should be smaller than the expected thickness. It may be difficult to develop validation rules for other types of unconformities.

Petrophysical data problems

Developing validation rules and screening petrophysical data for potential errors is, for the most part, a statistical exercise. The use of geostatistical methods for validation is discussed in more detail below, but statistical means and distributions are the keys to developing good validation rules for petrophysical data.

- As with other types of data, start the validation process by screening for data that cannot (or should not) occur naturally. For example, porosity values > 50% should always be considered suspect. Water saturations < 5% would also be very unusual. The DBM should work closely with the petrophysicist and/or log analyst to develop validation rules and ranges to screen out absurd values.
- In most regions, there are previous petrophysical data available that will help define the range of likely porosity and Sw values for most rock types and reservoirs that have already been drilled. Once the database has been cleaned up, it should be possible to refine these ranges even further using geostatistical methods.
- If the data storage format is decimal percent, any values reported over 1.0 are considered format errors. If the storage format is percent, any values less than 1.0 might have been entered as decimal percent by mistake and should be reviewed.
- Virtually all geotechnical data have a certain global range of possible values, beyond which the data are considered impossible. Again,

working with the petrophysicist to help define these validation rules will identify most of these possible errors.
- Identifying unreasonable data is very difficult. Unreasonable data are values for a particular attribute that can occur in nature, but would be considered unreasonable under local conditions. These validation rules require a fundamental understanding of the data, local conditions, and the attribute itself.

Directional survey data

Deviation survey data fall within basic geometric limits; developing validation rules for this type of data requires application of these limits.

Duplicate depth points should always be flagged as potential errors because most surveys are continuous with regular depth intervals. Also, a large gap in the depth values could indicate potential missing data or other errors.

In most cases, a validation rule can be developed to flag any *rapid or abnormal change* in hole inclination or azimuth. A very high dogleg angle for a single point is unlikely. When looking for rapid, abnormal changes in the hole azimuth, the absolute angle difference between two adjacent depth points may show an apparent rapid change. (From 358 to 002 is only 4 degrees in azimuth, but a simple computation might blindly report a change in azimuth of 356 degrees.)

In boreholes with no lateral sidetracks, any *very high-angle borehole data* should be flagged for follow-up. Negative drilling angles in nonhorizontal wells or drilling angles in excess of 90 degrees should always be flagged as potential errors.

USE OF GEOSTATISTICAL METHODS

Statistical methods of validation are very powerful tools when the database is large enough to be statistically meaningful. Even if the statistical analysis tools are not included as part of the DBMS, selected data can be exported, loaded to a spreadsheet or statistical application, and analyzed. The graphical and numerical methods discussed here can be the best means of developing effective validation rules.

Histograms and probability distributions

The simplest and most straightforward method of applying statistical methods is through the use of histograms and probability distributions. All the database values for a particular data element are plotted as distribution histograms and cumulative probability curves. From these graphical displays and the statistical parameters associated with them, it is possible to define the mean, minimum, and maximum points. From the probability plots it is possible to define the 10th percentile and 90th percentile (and other parameters) that show the most likely range of values. Using these points as a guide, data that fall near the 10th and 90th percentile points should be flagged for review. Figure 11–1 illustrates a typical distribution histogram with cumulative frequency curve used as part of the data validation process.

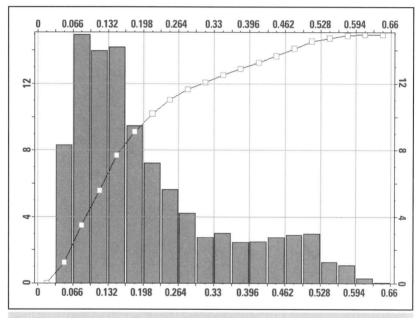

Fig. 11–1 Histogram and cumulative frequency curve for data quality control. Constructing this type of histogram for various appropriate data elements can be a powerful QC tool. Spurious data points can be isolated, unrealistic data recognized, and overall limits can be developed for future data validation checks.

Log data normalization procedures

Several authors have presented general techniques for normalizing log data measurements. In summary the application of this methodology for normalizing log data in a field or limited geographic area is as follows:

1. All data points for a particular curve in a specific, relatively narrow stratigraphic interval for multiple wells in an area are plotted on a histogram. This becomes the field reference for that curve in that interval.
2. Individual histograms for each well in the field are plotted for the same curve over the same interval at the same scale as the field reference histogram generated in the previous step.
3. Each individual well histogram is compared with the field reference histogram (by visually overlaying the two plots). Differences or shifts in the mean, maximum, and minimum are noted on each well. The curve data for that well are then shifted until the two histograms are closely aligned. Shift values (mean, minimum, and maximum) should be stored somewhere in the database.
4. After all wells have been shifted (normalized) to the field reference, the field reference histogram is reconstructed. The final field reference histogram is then used to normalize any new wells that are drilled in the field using the same procedure.

Regional statistics

Like distribution histograms, regional distributions of data can be used to identify trends in the data that can help construct validation rules. The most effective way of applying this technique is using map displays of the individual data elements.

When there are sufficient data available in an area (as is generally the case with field development data), it is possible to select appropriate data types and construct *regional or field-level trend maps*. These maps form the basis of visual validation of new data added to the database that shows significant variation from the regional trend. Of course, these sometimes subtle variations are how new prospects are identified and are not always indications of errors. Either way, flagging exceptions for further review can help identify actual data errors or legitimate exceptions that may indicate new opportunities.

Derivative maps, or trend residual maps, are a view of the data after the regional trends and structural surfaces have been flattened or removed. Anomalies on these maps show data points that are significantly greater or less than the expected regional trend. Again, these should be flagged for review as either data errors or potential opportunities.

Database tools

Most of the statistical analysis of the contents of a dataset can be done using tools and functions that are already built into the DBMS. While not always as effective or visual as graphical methods, they provide a rapid analysis that allows a quick look at the trends in the data.

Using SQL or other database query tools, it is possible to evaluate and retrieve data for quick review and further follow-up.

Most DBMS software has built-in functions that allow the DBM or DBA to quickly generate most basic statistical functions, including maximum, minimum, mean, and standard deviation. If there is sufficient justification, the graphical methods discussed earlier can then be used to further refine the statistics.

When special statistical analyses are needed or if some computational operation is required, it is sometimes best to create specialized user-defined functions (UDF's) for this purpose. As with other tools of this type, UDF's should be documented and added to a catalog in a shared area so that other users can employ them.

VALIDATION TOOLS

After a basic set of validation rules and workflow procedures have been developed and tested, it is best to create a standardized approach to the data validation process. This serves several purposes. First, it provides a structured and systematic approach to data validation that ensures all the proper checks and validation tests are performed. Second, developing validation tools helps automate the validation process. Finally, a formal validation tool embodies much of the expertise of the user and DBM that preserves the procedures, policies, guidelines, and standards for data validation.

Functions and tools

Many of the validation methods discussed above can be applied using internal DBMS functions and tools. If these internal functions are not available, it should be possible to export the data to a statistical analysis package or a spreadsheet. One drawback to this approach is that if any editing is done on the data during this process, the changes may not be reflected in the database.

Using this methodology requires very close coordination between edited data and the master database to ensure the edited data update the master data accurately.

Programmatic solutions

To apply a measure of automation to the validation process normally requires the assistance of a programmer to develop a validation tool. Unfortunately, applications developed specifically for data validation are not typical, off-the-shelf applications. The development of a data validation application must incorporate the validation rules, reporting, and auditing functions. As with any application, the workflow, requirements, and functionality should be well planned before any development work takes place.

DATA EDITING

Once data have been identified by the validation procedures as possible errors, there must be a mechanism in place so the user can make the editing changes or request the assistance of technical support personnel.

Editing methods and options

Allowing full, unrestricted editing privileges on database contents is a dangerous issue. If the user has sufficient security level and authority, it is possible that data will be changed without the knowledge or authorization of the database manager (DBM). The DBM can take two basic approaches with regard to editing.

First, the DBM can grant access to a database for individual users. This requires making several considerations:

- Should the user have editing access to anyone else's data? If there is no strong justification for this additional level of access, granting edit access to any data other than the user's is not advised.
- When users access the database, any editing work (including additions, deletions, and editing) should be stored in a separate archival database in case the original (pre-edit) data need to be recovered.

When the corporate resources are sufficient to support it, a centralized group of data loading experts and data entry operators should be established to assist with loading.

Data editing tracking and audit

Although data editing and clean-up are essential for the overall health of the database, the process involves a human interface and is therefore a potential source of errors in itself. During the editing process, it is important to maintain some mechanism for tracking what is changed, who is making the changes, and how the old (edited) records will be stored in some sort of recovery area (see *History Files and Deleted Records Files* section in chapter *Designing the Database*). In this way, if editing problems are encountered, it should be possible to recover the original (pre-edit) data and restore the database to the previous conditions.

Reporting data problems

Validation methods can identify potential data errors. However, the most likely source of error recognition is with the end-users of the data because they apply the information during interpretation. In either case, whether identified by a user or during routine validation processes, there needs to be a mechanism for reporting, tracking, and correcting problems.

Problem tracking database. One solution for reporting, tracking, and correcting data problems as they are identified is to use a centralized problem tracking system. Since we are dealing with database information, it is reasonable that the system should link to the underlying database where the data are stored. An example of a prototype problem reporting system interface is shown in Figure 11–2. The fundamental elements of this system should include the following:

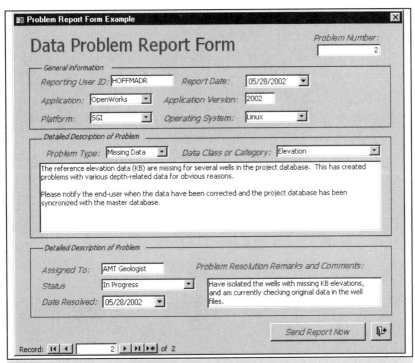

Fig. 11-2 Prototype of a Data Problem Tracking System Report Interface. The entire system would include various other options, functions, and features as discussed in the text.

- The problem should be described, including all information necessary to identify the possible error, including who identified it and what criteria were used for identification.
- Every potential error identified should be assigned to a specific individual for corrective action whenever possible. In most cases, the ownership information on the data record where the error is found identifies the most likely person to resolve the problem. If that person is unavailable, the appropriate asset team leader or group supervisor should be listed as the responsible party.
- Once the problem has been resolved, a short description of the corrective action should be stored in the database. This information can be very useful if there ever arises a need to review the audit trail of data editing to determine how, by whom, why, and when a particular data item was altered.

Data problem categories. Several general categories of data problem types need to be reported and tracked by this system.

Data format problems are generally identified by the validation process during loading or routine review of the data. The DBM in most cases may handle this class of data problem without consulting the data owner or user.

Data identified as being outside the normal and expected statistical range are possibly the result of data entry errors and poor validation rules. This class of error requires that the user provide the corrected values. The DBM should review and modify the validation rules as necessary to prevent the problem from recurring.

Routine data entry errors, especially those identified by the user as being anomalous when compared to regional trends, are mainly the responsibility of the user for corrections. Obviously, some empowerment must be given to users to update and correct their own data.

Mechanisms for tracking problems. The concept of a problem tracking system is fundamentally the same approach taken by most IT help desk groups. When problems are reported, they are assigned a tracking number, a responsible party for correction, and information about the problem. When the problem is resolved, the job ticket is closed and maintained in an archive for later audit and review. Several commercial systems (see Appendix A) are available that perform these functions and could be adapted to a DBMS environment.

Most help desk software can be delivered in one of two ways. A browser-enabled system is probably the best approach. Problems are reported and tracked via a Web-based interface, making it available to all users regardless of computing platform. This approach requires an underlying database to store and manage the problem reports, as well as an individual or group to maintain and run the system.

Another possible approach is to use the capabilities of company e-mail systems that are already in place. Both Lotus Notes® and MS-Exchange® mail systems provide the ability to build custom input forms and a mechanism for managing and monitoring the problem reports. Most users are familiar with the e-mail system and will be comfortable with this environment.

A phone-based system is not the ideal stand-alone approach, but having a telephone help desk front end interface to either the Web-enabled or mail-enabled system is sometimes a useful feature. However, a phone-based system requires additional dedicated personnel who may not be available.

DATA QUALITY CONTROL

Overall data quality control is the cornerstone of effective data management. Quality control requires the establishment of policies, standards, procedures, and guidelines (PSPG). These PSPGs must be documented and enforced to preserve and maintain the data quality. Finally, the PSPGs need to be reviewed and revised periodically to maintain currency with respect to changes in the database.

Quality control methods

Data quality control should be done as an ongoing process throughout the life cycle of the database. No large database, no matter how carefully it is controlled, can ever expect to be 100% accurate. While a 95%–98% level of accuracy should be the minimum acceptable level, there will always be room for improvement and minor editing. Quality control methods done using on-demand and systematic reviews are the key to maintaining the integrity of the data.

The on-demand validation checks can be triggered by a review of the statistics generated by the problem tracking system. When a large number of similar types of errors are reported or when a particular data type is consistently being edited, a review of that portion of the database may be justified. Based on this review, additional errors may be identified, and the validation rules can be revised to reduce the errors.

In addition to on-demand validation, there should be routinely scheduled screening of the database to ensure that every part of the database is reviewed and monitored at some point. In some cases, stored data will go unused for months or even years. If users are not accessing the data, errors may not be found until the information is needed (usually on a priority/rush basis), at which point there may not be enough time to do the necessary clean-up.

Quality assurance and documentation

The users of a geotechnical database need to be certain the information stored in the database is accurate, complete, and always available. If this level of confidence is not present, users will maintain personal databases and not utilize the master database system, defeating the purpose of having a DBMS.

How often the routine validation checks are made on the data depends on how active and how dynamic the database is. If the data contents are primarily static and archival in nature (not the most likely case), frequent validation checks are not justified. On the other hand, if there are dozens or hundreds of users accessing, editing, and revising the data, with new data being added daily, validation checks may be needed on a weekly or monthly basis.

Validation rules database. As stated in other parts of this text, documentation is a key element in the success of a data management system. The two main elements of this documentation include a database of validation rules and a thorough documentation of procedures. The complexity of geotechnical data and the sheer volumes of data involved make it essential to thoroughly document all validation rules for the database. For each data type, this documentation should include the following.

- The objective, scope, and purpose of the validation rule should be described in detail. It should be clear what the intended result of the validation is.
- For numerical data, the maximum and minimum allowable limits should be specified. Where appropriate, the units of the numerical data should also be specified.
- For text or date elements, any specific format requirement should be specified, including examples if the format is complex.
- When revisions are made to validation rules, a notation should be made that documents why the revision was justified, when the revision was made, and who made it.
- Criteria used for development of the validation rule are probably the most important part of the rules database. Every validation rule is based on criteria determined from local experience, general knowledge of the data, and/or statistical analysis of the contents of the database.

Data dictionaries. Earlier in this book, the subject of data dictionaries was discussed in detail (see *Designing the Database* chapter). Part of the data validation and quality control process requires that a thoroughly documented data dictionary be available during the validation review process.

Reviewing data validation methods

Whenever errors or potential errors are identified in the database, one of two possible situations has occurred; either the data are within the limits of current validation rules and were simply entered incorrectly or the current data validation rules are not sufficient to screen these errors during the data entry and editing processes. The second category is the more troublesome because little can be done to prevent someone from inputting incorrect values.

Quality assurance refers to the policies and activities that are conducted to ensure a defined level of data accuracy and quality. Quality control, the prevention of unwanted data errors in this case, is the result of proper quality assurance. Regularly scheduled reviews of the content of the database should be conducted as part of the overall quality control and quality assurance procedures. As part of these reviews, activity statistics can be generated that show where most of the data loading and editing activity occurs. These parts of the database should be scheduled for more frequent validation checks because this is where most of the new problems will enter the system.

For a dynamic database, a periodic review of database content distributions should be reviewed and compared with the current validation rules. As the data content changes, it may be necessary to revise the validation rules to account for the changes to the underlying data.

Modifications to data validation rules should be reviewed whenever there are a substantial number of errors identified with a data element or data type or when the statistics of the underlying data change substantially. When the validation rules change, the changes should be reflected in the validation rules database. The data for which rules apply should be rescanned with the new validation rules to ensure that the results meet expectations.

The most effective methods of data validation are those that are incorporated into the user interface. The next chapter discusses the process of planning, developing, and testing the user interface using many of the strategies presented in earlier chapters. Part of this planning and design process must include how data validation will be included in the interface.

12

Designing the User Interface

In most cases, commercial DBMS software includes some type of predefined user interface. For custom applications or customized versions of commercial software products, it is sometimes necessary to develop user interfaces that meet existing workflow requirements or specific needs and preferences of the end users. This chapter does not cover the programmatic aspects of developing a user interface but presents concepts and suggestions to make that process more effective and efficient.

USER INPUT AND FEEDBACK

The importance of user input and feedback as a critical part of a successful data management project has been stressed throughout this book. The user interface is probably the single most important part of this process because it is the point where the user interacts with the data. No matter how well planned and executed the other parts of the system are or how efficient the DBMS, if the interface is cumbersome, difficult to learn, and absent the desired functionality, the entire process will fail. If the clients (the users, in this case) are dissatisfied with the product, they won't use it.

Planning the interface

Careful planning at this stage is vital. Key users from various disciplines in the organization should be invited to participate in this process. The developer and DBM should meet with the user community in a series of information-gathering sessions to develop a general prototype of the interface.

Develop a survey strategy. Before meeting with any users, the DBM should meet with the developers to discuss what has been planned (and may already be implemented) regarding the objectives of the project, the types of data involved, data formats, and expected use. From this information, a series of interview questions can be developed to help focus individual users on the task of helping design an interface that will provide the best possible solution. Some typical issues that will need to be addressed are as follows.

- What general form will the interface take? If a map-based interface is required, how will the user select information from it? If a forms-based interface is all that is needed, how much detail will be required on the form?
- What are the expected levels of expertise? Will there be a complete spectrum of user experience, ranging from novice to expert, that will require modifications to the interface based on user expertise, or will a single interface be sufficient?
- What types of functionality are required? Does the user expect to have computational capabilities, or will the interface be used strictly for data browsing and reporting?
- How much data will be entered via the user interface? If the expected level of data entry is very high, the data entry personnel (if they are other than the actual users) should be included in the individual and possibly group interview sessions. Gaining a fundamental understanding of the workflow processes will be valuable in planning the interface.

Individual interview stage. After developing a focused list of interview questions, the DBM and developer conduct individual interviews with the end-users. These interviews should be short and focus only on the actual interface development. Depending on how broad the data management project is, users from all affected disciplines should be included in the interviews. Data entry personnel or technical assistants who use the interface should be included as well.

At this point, it may be necessary for the DBM and developer to meet with selected users to perform a limited workflow evaluation of what the users do and how they actually work with the data. A great deal can be learned by spending a few hours watching how a person conducts his routine interpretive work and listening to what he needs (or think he needs) from a data management product.

Preliminary interface design. Based on the results of the individual interviews and work flow analysis (as required), the DBM and developer should lay out (on paper) a general plan of what the interface should look like, what functionality should be included, and how the different parts of the interface should interrelate.

This step should be "prototyped" on paper, and should not require a working prototype. It is better to make mistakes and revisions on paper than waste valuable programming resources developing working code. The paper design should demonstrate conceptually what the interface should look like and, in general terms, how it should function.

Group interview stage. Once the conceptual design has been completed, the users who were interviewed individually should be assembled and the preliminary interface concept should be presented to them. It may be advantageous to deal with multidisciplinary teams or with groups of users in the same discipline depending on specific "corporate culture" or personnel issues.

Again, this should be a short, informal review session aimed at correcting any misconceptions, poor designs, or bad ideas before any substantial development time is invested in the project. It is much easier to correct a concept on paper than to correct an end product that already has a lot of time and manpower invested in development.

Prototype development stage. Using the feedback and suggestions from the group interview process, the DBM and developer should resolve any problems or design flaws; then the actual prototype development can begin. This may involve work by internal programmers or contract programmers, but the DBM and project developer should work closely with the programmers to ensure the conceptual design is accurately represented in the prototype.

Group evaluation and critique. Once a prototype has been developed, it should be tested using data sets that have been provided by the users in previous parts of the planning process. After this preliminary testing has been completed, the user group should be reconvened and the prototype rolled out and demonstrated. All feedback from this session, or series of sessions, should be provided via the developer to the programmers as the final product is developed and launched.

User critique and feedback

During the development phase and subsequent rollout when the product is actually deployed to the users, it is important to provide some mechanism for soliciting user feedback about the interface. The user community can provide valuable suggestions for improvement if there is a simple (and painless) process for providing critiques of the interface or overall database system.

One way to provide this feedback is to include a feedback option within the interface (at appropriate locations) that will allow the user to create comments and suggestions, report bugs, and request additional features. In most networked environments, this can be accomplished easily by building a mail-to function into the interface that will launch an e-mail message to the designer and DBM. These comments should be cataloged and prioritized for action, with appropriate notification sent to the users when the new features and enhancements have been added to the interface.

Using an intranet site (see *Role of the Database Manager* section in *Planning Database Projects* chapter) to inform users of the status of problems and solutions is a very effective way of keeping users in the loop.

INTERFACE DESIGN OPTIONS

When considering the design of a user interface, two basic directions can be followed. The older, conventional interface design uses a linear, structured approach that provides the user with lists of options that move the user through the interface in predefined paths. These typical mainframe interfaces are surprisingly widespread even today.

A more flexible interface can be designed using object-oriented programming methods that provide a much wider range of control, options, and customization. The interface preferred by most geoscience workers uses some type of map-based or GIS interface.

Conventional (menu-based) interface design

The conventional menu-based interface was designed in the early days of computing, long before the advent of current GUI designs and the Windows®

approach to computing. This was necessary because of the text-based display formats and lack of sophisticated GUI tools.

Functional features. Most early user interfaces for computers were developed long before the Windows® environment of GUIs. At that time, graphics were limited, and user interfaces generally consisted of a collection of individual screen displays that presented the user with a series of options as menus of choices. Selecting a menu option moved the user to another screen, and so on through the system.

The disadvantage of this approach was apparent even then—it was difficult or impossible for the user to deviate from the preset menu choices and move freely from one choice to another menu unless that path had been preprogrammed by the developer. Even with today's sophisticated GUI capabilities, many user interfaces still restrict the user to this limited design philosophy.

The following series of screenshots illustrates the menu-based user interface concept. The main menu (Fig. 12–1) allows the user to move the highlighted selection with the arrow keys or mouse to the desired operation. then the user presses the Enter key or clicks the right mouse button. Figure 12–2 shows the submenu screen selected from the main menu. Figure 12–3 shows the application selected from the submenu.

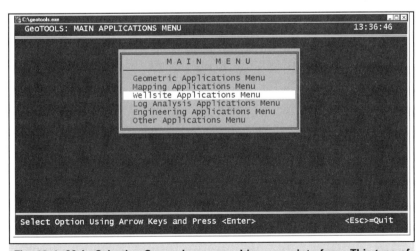

Fig. 12–1 Main Selection Screen in a menu-driven user interface. This type of interface was much more common in early, non-graphical computer systems, but is still used occasionally today. This example shows a number of geotechnical program group from which the user selects one category.

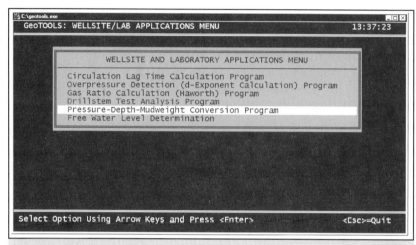

Fig. 12–2 Second-Level Menu in a menu-driven user interface. This menu is activated when the user selects a category from the main menu. In this example, the user has selected the "Wellsite/Lab Applications" category from the main menu, and the second-level menu shows all the programs within that category.

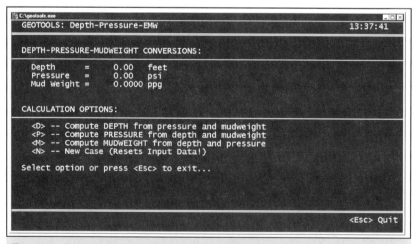

Fig. 12–3 Computation screen in a menu-driven user interface. The user has selected the "Depth-Pressure-Mudweight" program from the second-level menu, and the input screen for that program has been displayed.

The development of a menu-based interface is very straightforward once the fundamental workflow is identified. These interfaces are still used by mainframe interface designers where there is no GUI capability and for small utility-level applications where the interface is simple and uncomplicated. Because most large-scale database systems are more complex, the user interface must be more flexible and provide more options for the user. While a menu-based design might be a suitable starting point for planning purposes, a GUI design should be implemented wherever possible.

Object-oriented (form-based) GUI design

As computing power and graphical capabilities have improved, various Windows®-based application development software programs have become available. High-level, fourth-generation programming languages (4GL) allow rapid development and deployment of applications, and are the basis for most GUI applications in use today. Object-oriented programming (OOP) languages let the developer create graphical objects such as buttons, checkboxes, lists, and data entry fields. These objects are collected into forms—frame-bounded windows that can function independently of one another.

User interfaces that utilize this technology have many features and advantages. One advantage is the use of modular and reusable programming components. As a developer creates an object-oriented interface, the individual objects or components are effectively stand-alone bits of programming code related to other objects in the interface. The big advantage of OOP is that these individual modular objects can be reused in other applications with only minor changes. This speeds the development and deployment process and eliminates many of the coding errors introduced using conventional programming methods.

A user interface developed using object-oriented GUI objects also provides the user with a great deal of flexibility. This is not to say that all GUI applications provide this flexibility. Poor design and development can offset any advantages. The biggest advantage of a properly designed GUI interface is that users can move from one part of the interface to another according to their needs or workflow requirements and not according to a predefined set of menus. The best comparison is the difference between placing a telephone call directly to a person's extension rather than wading through six or eight levels of automated phone menus before getting frustrated and pressing zero.

In addition to the reusable component nature of GUI design techniques, development can be made even easier using off-the-shelf objects developed by third parties. These object or component libraries provide most of the routine and even very sophisticated solutions in prepackaged objects that can be used directly by the developer. After development, maintenance of object-based interfaces is somewhat easier than conventional menu-driven designs. When a particular object in the interface fails to produce the desired results, debugging and modification are isolated to that component without affecting other parts of the application.

GIS (map-based) interfaces

Most geoscience data are spatial in nature and are therefore well suited for a map-based user interface. A map-based or GIS application can display spatial data and allow much better visualization of the information. In addition, if the proper database links have been created, it is possible to "click on" or select features on the map-based or GIS interface and view / edit the underlying data. The conversion of this "ultimate interface" from demo to reality requires considerable preparation and effort.

Spatial metadata. Before geotechnical data can be viewed through a GIS interface, a set of spatial metadata must be created. Metadata is the information about the data and not the actual data itself. For example, the locations of the corner coordinates of a geological map are a form of metadata about the map. Using the metadata, a shape representing the boundaries of the map can be displayed on a GIS interface. With the proper links to the actual data, the user can select the shape; the data are accessed from the database and displayed. In most cases, the software vendor promoting a map-based GIS system will downplay the fact that the client has to provide the metadata before the system is fully functional.

GIS considerations. When considering the development of a map-based GIS interface, keep the following considerations in mind.

- Who will be responsible for preparing and loading the metadata? If the client company does not have the in-house staff to do this work, the cost of contracting this work to a third party and long-term maintenance issues need to be addressed.
- Who will load the spatial data? Loading the actual spatial coordinate data requires a great number of conversions and internal consistency

checks. Someone with extensive experience working with projection systems and spatial datasets is required for this work.
- What capabilities does the system have for importing and exporting data? While it is nice to be able to display and browse data on a GIS interface, if the system does not have the capability to link directly to the existing or planned DMBS, the information shown on the map will have to be loaded to the GIS system's internal database. This creates a static, unsynchronized dataset that does not truly represent the actual content of the database. Furthermore, if the GIS system cannot export data to an interpretive application, it is of limited usefulness.
- What are the system capabilities for adding functionality (i.e., programming extensions to the standard product)? Most GIS products are generic and can display virtually any spatial data. Some of the requirements and data types used by geoscientists require that some custom extensions to the functionality of the system be developed. How easily this is done will have great impact on future maintenance and modifications.
- What programming languages are used to develop customizations? If customization is required by the user or for specific operational needs, evaluate a programming language(s) that may be available for the modifications. A proprietary language supported only by the vendor may be a drawback, in that it locks the client into using that vendor for all programming and limits the ability to utilize code modules developed by other vendors. If the programming language is more standard (Visual Basic or C++), you still need to determine if in-house staff will be available when modifications are needed.

Remember to insist on a product demonstration with real data. Demo datasets are carefully crafted and tested to showcase the features and capabilities of the system without problems or limitations. If you are seriously considering a GIS interface, insist that the vendor load and demonstrate the system with samples of corporate data that represent all the types of data to be accessed or managed by the GIS system. The pilot datasets used in the planning process can be provided to the vendor for demonstration purposes, assuming proprietary information has been removed or modified to preserve the confidentiality.

Validation Considerations

While developing the interface, pay close attention to data validation issues. The user interface is the "front line" of the data collection and editing process; as such, it is the primary point of data validation. If the validity checks, data ranges, and other data QC methods are "built-in" to the interface, most of the major data errors will be eliminated before they enter the database.

Form-based validation

The interactive portion of the user interface will generally consist of various forms that are displayed on the user's workstation. Regardless of whether these forms are displayed on a PC, a UNIX workstation, or via a Web browser, this is the point in the data management process where data are entered, edited, and deleted. Therefore, focus your attention and planning on this portion of the interface. Depending on the type of data being handled by the system, there are various useful techniques for providing form-based data validation. An example of a form-based interface that illustrates various common features is shown in Figure 12–4.

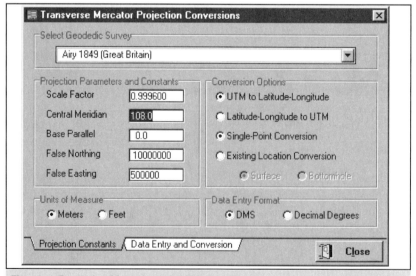

Fig. 12–4 Example of form-based validation. This form includes most of the common "controls" that are used in GUI forms. Each of these controls is discussed in more detail in the following sections.

Validation ranges. When an interactive form is being constructed, the developer uses a variety of controls or objects that can be used to display, enter, and/or edit data. Each of these objects has specific attributes that can be set to check for data that meet minimum or maximum values or that conform to specific input formats or templates. In addition, it is possible in most cases to provide a default value and display warning messages when the user attempts to enter or edit data beyond the preset limits.

Drop-down and list boxes. Anyone who has used computers for even short periods is familiar with the concept of drop-down boxes and list boxes (even if they didn't know what they were called). The *drop-down box* is a small display field that can be expanded by clicking on an arrow button next to the box, as illustrated by figure 12–5. When this is done, a longer list of values is displayed (see Fig. 12–6), and the user can select one and only one of the values from the list.

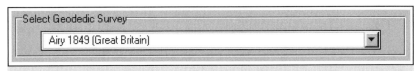

Fig. 12–5 Drop-down list box (closed position). When closed, this box shows only the current value for that data field, and takes up very little room on the form. The "button" on the right side of the box activates the drop-down feature.

Fig. 12–6 Drop-down list box with list activated. When the user "clicks" on the button on the right side of the list box, the drop-down list opens. The user can then select another value from the list. When a value from the list is selected, the drop-down box closes and the newly selected value is displayed in the box.

A *list box* is effectively the same thing, except the longer list of values is shown in a fixed-size display area (usually with a scroll bar that allows the user to move through the list). With list boxes, the user can select one or more of the values from the list but is prevented from entering any other value in the field. This is an excellent and generally foolproof method of data validation when the data in the box are relatively static and the user *must* select one or more options from a predefined list.

Pick lists. The *pick list control* combines the basic elements of the list box, but includes two "side-by-side" list boxes. Between the boxes are commonly a row of buttons with ">" or ">>" arrows that allow the user to pick one or more items in either list and move the selected items from one list to the other. Figure 12–7 shows an example of a pick list interface control.

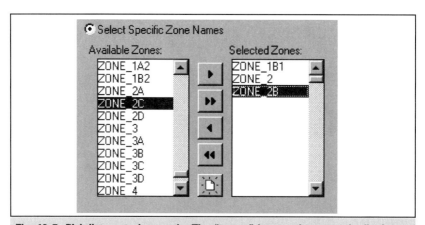

Fig. 12–7 Pick list control example. The "arrow" buttons between the list boxes allow the user to move one (selected) item from one box to the other. Other buttons allow all the items to be moved, or to "reset" the boxes to their original state.

The pick list is commonly used when the user wants to select several, possibly noncontiguous items in a list to build a subset of the original list (such as a subset of formation tops, as in the above example). In most cases, the information contained in the left pick list is created from the database as the form is loaded.

Combo boxes. A variation on the drop-down list box is the combo box (the actual control is a combination of a list box and a data entry field, but technically the control is referred to as a "combo box"). This control looks exactly like the drop-down list box and functions in the same way with one important exception: users can enter a new value into the main part of the box if they don't find a choice in the list. This provides a great deal of flexibility to the user, but it does raise validation issues.

Radio buttons. Radio button controls consist of a series of two or more buttons from which the user selects one. Only one button can be selected at the exclusion of all the others.

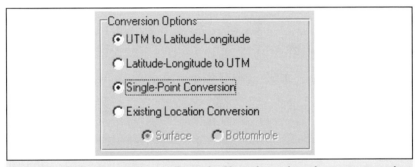

Fig. 12–8 Radio Button Control Example. Note that sub-options are not active unless the main level option is selected. With this type of control, only one selection can be active at a time.

This type of control should be used when the choices are static (because they are built into the form design), and the user should only be allowed one choice from those presented. The radio button approach is similar to the drop-down list box in that only one choice is permitted from a series of choices. In this case, however, the options are designed into the form (while the drop-down list can be created from information in the database). Default values can be used with radio buttons to preselect the most likely choice for the user.

Checkboxes. The checkbox control is a collection of options from which the user can select any or all of the choices. These controls should be used where more than one choice will be allowed. Again, however, the choices are static and don't change (unless the form is modified). This type of control is similar to a multiple-choice list box where the user can choose more than one option if desired.

Checkboxes are helpful when there are multiple default values, which can be preselected when the form is loaded.

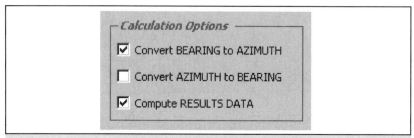

Fig. 12–9 Example of Checkbox GUI Control. This control allows the user to select one or more (or no) options from a predefined list of choices.

General text and memo fields. Text and memo fields are the most difficult type of control to use when input/editing validation is required. Although a certain amount of validation can be included as input formats or templates, much of the data validation in these fields must be handled outside of the form environment (see previous chapter). Where a limited amount of text is allowed, the text field format should be used. If a variable amount of text-based data is allowed, the memo format field is a much better choice.

Numeric fields. Numeric fields allow numeric data only. Input can be validated to determine if the input/editing values fall within a certain range of values. In addition, predefined input templates can enforce specific input formats. For example, if a data field for porosity data is included on a form, the input validation format can be set to reject any values greater than 0.60 or less than 0.00. This guarantees that the data will not be percentages and will be realistic.

Spinboxes. A more rigorous method of controlling numeric data entry is by using spinbox controls. These form controls generally have a default value in an entry field and adjacent up and down arrow buttons. Pressing the up or down buttons increases or decreases the value in the input box by a predefined amount (Fig 12–10).

Fig. 12–10 Spinbox Control Example. This type of control generally allows the user to increase or decrease the value shown by preset increments using the "up" or "down" arrow on the right side of the control. In some cases, the user can directly enter an amount from the keyboard without using the increment or decrement buttons.

The developer has complete control over what the default value should be, the upper and lower limits of the field, and the increment value. This specialized type of control should be used when the input/editing data are within known limits and the amount the value should be incremented is known.

Date controls. There are two common methods of entering dates using form controls. The simplest type allows the user to input a date but only allows a particular date format. Be careful, though: it is very easy to confuse dates that are not intuitive.

The best approach does not allow any data entry but lets the user select a date from a drop-down (or pop-up) graphical calendar. This approach is the easiest when date validation is important.

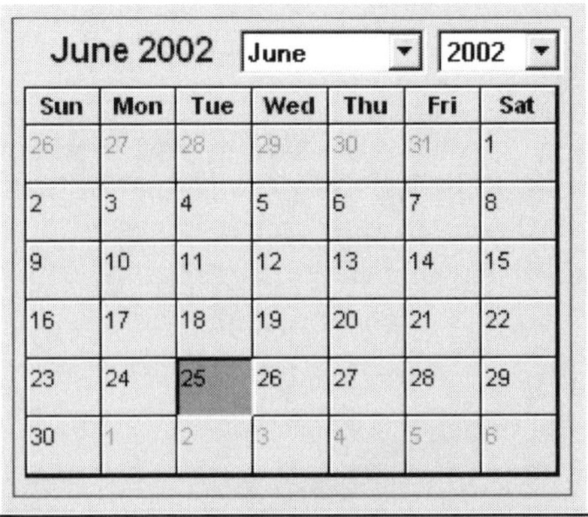

Fig. 12–11 Example of GUI Calendar Control for Date Entry. This is a very flexible control for entering or selecting calendar dates. There are many options and variations on this type of control, but most versions will include options to select the year and month, and then select a day from the "virtual" calendar page.

Table-based validation

The use of table-based validation control is slightly less interactive than form-based validation because the validation rules are built into the actual table structure. In most cases, the level of validation control at the table level does not exceed that which can be accomplished by building the rules into an interface form. Because of this, many developers rely solely on form-based validation and do not validate at the table level.

Unless there is no possible way for someone to access the data tables directly, the validation rules should be included at the table level as well. This will protect the data integrity from the well-meaning (or even not-so-well-meaning) user who accesses the data directly through outside links without passing through the form-based interface.

Import and export considerations

The most difficult part of data validation occurs when data are imported into the database from other sources. During the import process, most systems perform minimal validation as the information is read and stored into the data tables. Commonly, data from legacy database systems are exported into ASCII-format text flat files for subsequent import into the new database. This further complicates the issue because the data in the flat files may need to be imported to multiple tables in the new database. At that point, the relations between tables, key index fields, and duplicate data are concerns.

To minimize the impact of these problems, the data files that need to be imported should be defined (by the originating database) in such a way that there is a close correspondence between the format and content of the input file and the structure and validation rules of the receiving database. Further checks on the validity of the imported data may require programmatic solutions. Exporting data from the database is somewhat easier because the validation rules have presumably been applied to the content of the originating database. From the perspective of the database or application that will ultimately receive the data, however, the same issues discussed earlier should be considered.

Programmatic solutions

Even with the best table- and form-based validation and methodical import processes, there will be instances that require *ad hoc* or predefined solutions with some degree of programming. Most data management systems have various script and macro development environments as built-in tools, which require little or no formal programming support. When possible, these tools can be developed and used by the DBM to perform data validation operations. Simple validation checks can be done using direct queries (with SQL statements), as discussed in the chapter on data normalization. For more complex solutions, developer or programmer support may be required. In these cases the DBM and the users should work together to develop tools that meet the overall requirements and objectives of both the data and the database.

IMPORTANCE OF USER INVOLVEMENT

As with all the other phases in the development of a DBMS, involvement of the end user is extremely important at this point. The goal of the user interface is to make the database accessible and useful to the people who desperately need to get to the data: the users. Creating a slick, complicated, graphics-rich interface may be exciting and challenging to the developers, but if it isn't functional, practical, and provides a solution to the user's data management requirements, it will never be used.

As discussed in the chapter on *Planning Database Projects*, meeting with the users again to better understand the actual workflow processes is the key to a successful user interface. It may be necessary (and is highly recommended) for the DBM and/or developer to sit in as observers during a user's day-to-day activities. This allows the users the opportunity to demonstrate what they actually do during a typical analysis and interpretation workflow, point out where data management needs are not being met, and explain why certain requirements and functionality will be required of the interface. These sessions can also be an opportunity for the DBM and developers to better understand the importance of the underlying data and the potential improvements that can be made, resulting in a better overall understanding of the entire workflow process.

During meetings and work sessions with the users, there should be a careful distinction made between what features of the interface are actually required to get the job done and which features and functionality could be classified as optional. The development of the interface should focus initially on the required features, but during the interface design, the additional functionality should be planned so the enhancements can be seamlessly added to improved versions of the interface later.

CUSTOMIZING COMMERCIAL INTERFACES

Not all data management projects are developed from the ground up. More commonly, there is some sort of existing DBMS that may or may not have a user interface already developed. In these cases, additional steps must be considered in conjunction with the preceding discussion.

Support and maintenance

As with customization of data models, changes to vendor-supplied interface applications will present future support and maintenance issues. Some commercial products in fact do not allow any customization. There are basically two main solutions to this problem.

- Use custom extensions. Many database interface products (for example, ESRI's ArcView software) allow the development and installation of custom modules written expressly for (or in some cases, by) the client that utilize the main product's library of functions and utilities. In many cases, these custom extensions enhance the functionality of the product, and can be of benefit to other clients. It is therefore in the vendor's best interests to include these enhancements in future releases and upgrades of the main product whenever possible.

- Partner with an application vendor. This approach assumes the software vendor is flexible and willing to work with the client to develop custom modifications to the user interface. In many instances, the enhancements are features that the vendor may consider adding to future releases of the software. By working with the vendor at the early stages of these customizations, the client company can have a great deal of influence in how the future production releases will be structured. While this approach may not be practical for many smaller companies, the concept is fundamentally the same for any size organization. Ideas are ideas, regardless of the size of the contribution.

At this point, the database has been designed, data have been loaded, normalized, and quality checked, and a user interface has been developed. The data management project will now move into a maintenance and improvement phase where the system will be "tuned" and modified in small increments. By reviewing the successes and mistakes of the project, adjustments and changes should be fairly routine. The next chapter summarizes the key points of this text to reinforce the most critical concepts.

13

Summary

This chapter summarizes the key points discussed in detail in earlier chapters in a checklist format. The intent is to recap the concepts and important points in a summary fashion that can be used as reference during project planning.

STANDARDIZATION

Project Standards

- ❑ General project standards
  - ❑ Define and document early in the life cycle of the project.
  - ❑ Use as a working "blueprint" for the remainder of the project.
- ❑ Staff requirements
  - ❑ Project manager
    - ❑ Practical, working background in geoscience field
    - ❑ Must possess fundamental knowledge of data management
    - ❑ Some information technology experience
  - ❑ Other project staff
    - ❑ Technical support staff (information technology background)
    - ❑ Clerical support (for data entry and editing)

- Personnel roles and responsibilities
  - Individual staff roles and responsibilities defined clearly
  - Periodic reviews, with adjustments as needed
- Project goals and objectives
  - Define and document overall project goals and objectives
  - Modify objectives as project develops and evolves

Documentation Standards

- Define documentation standards early in project life
- Documentation format definitions
  - Define standard format template, distribute to all staff members
- Document delivery mechanism
  - Define method for document delivery (Web, e-mail, etc.)
  - Set update and review schedule for document archive
  - Set access and edit authorities to specific staff members
- Documentation responsibilities
  - Define staff responsibilities for documentation
  - Set schedule for periodic reviews and revisions

Database System Standards

- Establish and document software standards
  - Nonstandard software must meet strict compatibility requirements
- Establish and document hardware standards
- Operating system standards, where appropriate

Data Format Standards

- Published standards should be enforced whenever possible
- In-house or proprietary standards can be applied in some cases
  - Determine that standard formats available are not appropriate for the data
  - Document and justify the creation of a nonstandard format

Interface Standards

- ❑ Define the general interface design and implementation methods
- ❑ Define development software, interface format, and accessibility options
- ❑ Establish basic functionality requirements with user input

USER INPUT AND FEEDBACK

User Surveys

- ❑ Predevelopment survey to determine user needs and requirements
- ❑ Determine real-world data workflows and interpretive processes
- ❑ Observation of actual interpretive work sessions to augment surveys
- ❑ Establish data management requirement baseline
- ❑ Conduct post-development surveys to measure success of project

Special Advisory Groups

- ❑ Nomenclature advisory groups
    - ❑ Stratigraphic nomenclature advisory group
    - ❑ Other "expert" advisory groups
- ❑ Data model advisory group to monitory changes to data model
- ❑ Interface design modification advisory group

Follow-Up Surveys

- ❑ User satisfaction survey to determine level of user acceptance
    - ❑ Allows comparison with predevelopment survey
    - ❑ Provides direct comparison with "before" and "after" conditions
- ❑ Project critiques.

Ongoing Communications

- ❏ Provide effective communication among project participants and users
    - ❏ Website or intranet
    - ❏ Help desk type of software
    - ❏ Problem reporting database

DOCUMENTATION

Project Objectives

- ❏ Documentation of design parameters
- ❏ Functionality documentation
- ❏ Interface design documentation

Preplanning Documentation

- ❏ Define specific user needs and requirements
- ❏ Define data types and formatting requirements
- ❏ Define data transfer and utilization requirements
- ❏ Document DBMS requirements

Database Design

- ❏ Provide data dictionary in reference format
- ❏ Provide for data model documentation

User's Operational Guide

- ❏ General operating manual for most users
- ❏ Developer's reference guide
- ❏ Training tutorials and sample data sets (where appropriate)

Appendix A—
Additional Resources

The published and Internet-based resources presented in this section are not a comprehensive set of data resources but rather are a starting point for further reading, reference, and exploration. Each reference or Web site is selected on the basis of relevance or usefulness and is provided with a brief review of the key points as they relate to the data management concepts presented in this book. The reader is encouraged to explore these starting points and add to the list based on personal interests and discoveries.

Note: Due to the fluid and dynamic nature of the Worldwide Web, some links presented here may have changed, moved, or been eliminated.

Data Model Information

Petroleum Open Software Corporation (POSC) This organization, a consortium of petroleum industry participants, provides an excellent starting point for more information on the POSC database standards.	http://www.posc.org
Petroleum Public Data Model (PPDM) The PPDM group is also comprised of a consortium of petroleum industry and software vendor sponsors.	http://www.ppdm.org

Log ASCII Standard (LAS) Information

General information about the Log ASCII Standard (LAS) for log data exchange	http://www.cwls.org/las_info.htm
New features in LAS 2.0 format	http://www.cwls.org/whatsnew.txt
Information about the download file	http://www.cwls.org/readme.txt
Download file with LAS utilities	http://www.cwls.org/lasaug95.exe
Information specific to the LAS 3.0 format	http://www.cwls.org/LAS_3_File_Structure.PDF
LAS 3.0 Sample files	http://www.cwls.org/LAS_30a.zip
LAS 3.0 File Certification Kit download	http://www.cwls.org/lascertify_kit.zip

Data Format and Standards Information

International Organization for Standards (ISO) This organization provides a paid service to supply various industries with recommended standards and business rules.	http://www.iso.ch/index.html
Canadian Well Logging Society (CWLS)	http://www.cwls.org
Society of Exploration Geophysicists (SEG)	http://www.seg.org
American Association of Petroleum Geologists (AAPG)	http://www.aapg.org
Spatial Metadata and Content Standards. The Federal Geographic Data Committee (FDGC)	http://www.fdgc.gov/metadata/constan.html

Sources of Public and Private Geotechnical Data

EOS Data Gateway The Earth Observing System (EOS) Data Gateway is a clearinghouse and search service for various NASA source data	http://redhook.gsfc.nasa.gov/~ims www/pub/imswelcome/plain.html
IHS Energy	http://www.ihsenergy.com

Sources for Geoscience Application Software

Schlumberger GeoQuest	http://www.geoquest.com/pub/index.html
Landmark Graphics Corporation (a Halliburton Company)	http://www.lgc.com/Landmark Portal/home/index.html
Society of Professional Well Log Analysts (SPWLA) The US-based organization for professional log analysts and petrophysicists. Their website maintains a comprehensive listing of petrophysical software, with direct links to all the major petrophysical and related software vendors.	http://www.spwla.org/pages/software.html

Relational Database Software

dBase Incorporated
The original dBase® product was developed by Aston-Tate Corporation, later acquired by Borland, and is now supported and maintained by an independent developer, dBase Inc.
http://www.dbase.com

Oracle Corporation
The original Oracle® RDBMS product was developed in 1977, and now Oracle® Corporation is among the largest software companies in the world. The Oracle® RDBMS product is the industry leader for information management software.
http://www.oracle.com

International Business Machines (IBM)
IBM originally developed the DB2® relational database product for use on large mainframe platforms. This RDBMS is still considered the "industrial strength" DBMS, and is now available on virtually every platform from mainframes to handhelds.
http://www-4.ibm.com/software/data/db2

Informix
Informix Sofware has been developing DBMS tools for 2 decades, and is a leader in data warehousing and Internet-based e-commerce.
http://www.informix.com/informix/products/productlist.html

Sybase, Inc.
Original developer of the SQL Server product. This product is no longer available, but some information is still archived by Sybase and can be found at this Web Link
http://www.sybase.com/products/archivedproducts/sqlserver

Other Related Software

Remedy Corporation
The Remedy Corporation makes a widely used "help desk" software application that may be applicable to a data management problem reporting and tracking database.
http://www.remedy.com

Oilfield Systems Limited
The DAEX (Data Exchange) software available from Oilfield Systems Ltd. provides a customized method of linking virtually any software product with any data management system.
http://www.oilfield-systems.com

World Wide Web References

World Wide Web Consortium (W3C) The W3C organization maintains specifications for most Web access and communication protocols.	http://www.w3.org
HTTP (HyperText Transfer Protocol) Specifications	http://www.w3.org/Protocols
HTML (HyperText Markup Language) Specifications	http://www.w3.org/MarkUp
XML (extensible Markup Language) Specifications	http://www.w3.org/XML

Development Language Reference Links

Java Developed by Sun Microsystems as a platform-independent web development language.	http://java.sun.com
Visual Basic Developed by Microsoft Corporation, Visual Basic® applications are ubiquitous on PC platforms running MS-DOS operating systems.	http://msdn.microsoft.com/vbasic
Perl A public Web development language that is used extensively by Web developers and administrators.	http://www.perl.com/pub http://www.perl.com/pub/v/documentation

Appendix B—Checklist for Geological Data Types

During the planning phase of a geotechnical data management project, a survey must be conducted to determine the types of geological data available, the formats to be accounted for, and the possible exceptions. The checklist presented provides a starting point for the DBM when conducting surveys and interviews with end-users. A similar checklist should be developed for geophysical data where needed.

BOREHOLE DATA TYPES

Surface Location Data Types

- Spatial Data
    - Well location coordinates in latitude-longitude and UTM
    - Well elevations (and units of measure)
- Hardcopy Data
    - Survey plat (hardcopy)
    - Survey coordinates (digital)
    - GPS coordinates (latitude, longitude, elevation)

Drilling-Related Data Types

- ❏ Cuttings Samples
    - ❏ Physical cuttings samples (bagged)
    - ❏ Cuttings description logs
- ❏ Drilling Time (Penetration Rate)
    - ❏ Curve-oriented data (digital)
    - ❏ Drilling time logs (hardcopy)
- ❏ MWD Data Types
    - ❏ Formation properties data (GR, resistivity, etc.)
    - ❏ Directional survey data from MWD tools
- ❏ Mud Gas Chromatography
    - ❏ Mud gas data on strip logs (hardcopy)
    - ❏ Mud gas analyses (hardcopy)
    - ❏ Raw mud gas data (digital)
    - ❏ Computed mud-gas ratios and other analyses (digital)
- ❏ Formation Tops
    - ❏ Predrill prognosis tops
    - ❏ Tops picked from drill time and cuttings
- ❏ Core Data
    - ❏ Conventional core material (physical)
    - ❏ Sidewall core samples (physical)
    - ❏ Rotary sidewall core samples (physical)
    - ❏ Core plug samples (physical)
    - ❏ Visual descriptions of all core sample types
    - ❏ Core photographs
    - ❏ Petrographic thin sections from core material
    - ❏ Core gamma-ray logs
    - ❏ Core tomography
- ❏ Paleontological Data
    - ❏ Faunal and age data from cuttings samples

Wireline-Related Data Types

- ❏ Directional Survey Data
    - ❏ Gyroscopic directional surveys
    - ❏ Conventional directional surveys
    - ❏ Borehole geometry logs
- ❏ Conventional Wireline Logs
    - ❏ Open-hole logging data
    - ❏ Cased-hole logging data
    - ❏ Formation testing data (wireline)
- ❏ Array Data and Borehole Imaging Data
- ❏ Borehole Geophysical Data
    - ❏ Conventional seismic checkshot surveys
    - ❏ Vertical seismic profile (VSP) data

INTERPRETATION DATA TYPES

General Interpretation Data

- ❏ Multidisciplinary Display Montages
    - ❏ Hardcopy montage
    - ❏ Digital data files used to create montage
- ❏ Maps and Cross-sections
    - ❏ Hardcopy maps
    - ❏ Multiwell cross-sections
    - ❏ Digital data files used to create maps and cross-sections
- ❏ Reference Logs (hardcopy)

Project Data Files

- ❏ Geological Project Data Files
    - ❏ Well-log data
    - ❏ General well data

- Geophysical Project Data Files
    - Seismic trace data
    - Velocity data
    - Statics analysis
    - Synthetics, checkshots, and VSP data
- Petrophysical Project Data Files
- Other Project Datasets

Formation Tops Data Sources

- Wireline Log Tops
- Core Data Tops
- Paleontological Data Tops and Intervals
- Seismic Data Tops and Markers

Core Analysis Data

- Conventional Core Analyses (Whole Core and Plug Samples)
- Visual Core Description Information (Hardcopy)
- Special Core Analysis Data
    - Point data (digital and hardcopy)
    - Graphical data (capillary pressure curves, etc.)
- Core Tomography Analysis

Mud Gas Analysis Data

- Computed Gas Ratio Data and Interpretations
    - Curve-oriented data (digital and hardcopy)
    - Graphical data (hardcopy crossplots)

Subsurface Interpretations

- Multiwell Cross-Section Interpretations
- Subsurface Structure Mapping

- ❏ Lithology Data
    - ❏ Lithofacies from core description data
    - ❏ Lithofacies from seismic attribute interpretation
    - ❏ Electrofacies determined from wireline logs

Petrographic Interpretations

- ❏ Thin-Section Analyses
    - ❏ Outcrop sample thin sections
    - ❏ Well cuttings thin sections
    - ❏ Thin sections from sidewall and conventional core samples.
- ❏ Thin-Section Photography
- ❏ SEM Photographs and Analyses
- ❏ X-Ray Analysis

Petrophysical Interpretations

- ❏ Conventional Analyses
    - ❏ Porosity, permeability, fluid saturations, etc. (point and curve data)
    - ❏ Crossplot interpretations (hardcopy)
- ❏ Multiwell Petrophysical Analyses
    - ❏ Electrofacies and petrophysical properties maps

SURFACE DATA TYPES

Physical Data Types

- ❏ Outcrop Sample Data
    - ❏ Physical samples
    - ❏ Visual descriptions
    - ❏ Petrographic thin sections
    - ❏ Outcrop photographs

Map Related Data Types

- Field Mapping Data
    - Spatial data for field samples and survey points
    - Structural mapping data
    - Outcrop orientation data
    - Lithofacies distribution mapping
- Topographic Maps and Elevation Survey Data

Remote Sensing Data Types

- Photography
    - Conventional and high-altitude aerial photography
    - Low-altitude oblique photography
    - Infared and other nonvisible band photography
- Satellite Imaging
    - Multiband spectral imaging
    - Satellite-based SAR and SLAR
- Nonorbital Imaging
    - Synthetic-aperture Radar (SAR) and Side-looking Airborne Radar (SLAR)
    - SPOT imaging
- Gravity Surveys
- Magnetic Surveys
- Other Special Surveys

Glossary

Due to the complex nature of the tools and applications used in database management, there are many discipline-specific terms, phrases, and acronyms that can be confusing and confounding to the new or casual user. The informal and "unofficial" definitions presented in this glossary should provide a working knowledge base of many of the important terms for the casual and experienced user alike.

4GL. see Fourth-Generation Language

aggregate function. Any computational method which acts on a multiple records and produces an aggregation result. Examples of this type of function are SUM( ), AVERAGE( ), MAX( ), etc. Aggreagate functions can be either built-in functions (see which), or user-defined functions (see which).

ASCII. Acronym for the American Standards Committee for Information Interchange. This standard has been in use for decades, and assigns specific code numbers for every printable text character (standard ASCII) as well as numerous graphical and non-printable characters (extended ASCII), such as printer command codes.

attribute. An attribute is a data field represented by a column within a database object. The database object can be a table, a view, or a report.

audit trail. A record of actions taken on any part of a data management system which allow tracking of all critcal information needed to reconstruct a series of actions. An audit trail can track changes to table structures, data type definitions, data elements, deleted records, edited or added records, etc.

BLOB. Acronym for a Binary Large Object. In most cases these are large image or vector files which are stored in the data table in their original, unaltered format. Curve, trace or other vector data is commonly stored in BLOB fields in data tables.

browser. Any software application that allows a user to read HTML content. Examples include MS Explorer, Mosaic, and Netscape.

built-in function. A simple, limited scope set of programming instructions that are provided by the vendor of the application or development software. Common examples of built-in fuctions perform trigonometric functions (COS, SIN, etc.) or aggregate functions (SUM, AVERAGE, etc.). See also user functions, aggregate functions.

business rules. The data validation and quality checking standards that are imposed on a database.

CASE tools. Computer Aided Software Engineering tools. These are various software applications which aid in database planning, design, and maintenance.

cultural data. Data pertaining to the man-made features in a mapping area. This could include rivers, roads, cities, etc.

currency. The degree to which the information portrayed reflects the current up-to-date status of the data in the database. When project data are created from a master database, the data are no longer considered current.

custodian. The person or group responsible for maintaining and managing the data type effectively and efficiently to support exploration objectives.

data. 1) The uninterpreted elements from which information is produced or derived through interpretation of the data. 2) A representation of specific attribute or aspect of a physical object (the object itself, or a representation of the object), or some measurement of the object's properties.

data (resource) management. 1) The documentation and coordination of the establishment and implementation of policies, guidelines, procedures & standards 2) A centralized function or group in a company whose task is to effectively manage company data as a critical corporate asset.

data administration. A centralized function or group responsible for the design and maintenance of the physical databases, application program, interface modules, associated design procedures and standards, and the physical security of company data, back-up, and recovery procedures.

data dictionary. A database which contains information about the data and data structures. This is also a catalog containing all the information about the tables, field structures, and other metadata about the database.

data element. This is the simpliest and most elementary unit of data that cannot be subdivided. These data elements are usually identified and described in a data dictionary.

data entry. The physical or electronic process by which information is loaded to a database system. In many cases this is a manual process involving keyboard entry of data, but may involve bulk loading of data from another database or file(s).

data flow. A diagram representing the use of data by business functions or processes, including information about the custody of the data at any point in the flow.

data independence. A technique used in data management that puts any data that can be changed into the database itself rather than "hard-coded" into a form or other program code. Values for list boxes stored in a table are independent, but not if those same values are stored in the program code of the form.

data mart. A data mart is a decision support system incorporating a subset of the enterprise's data focused on specific functions or actvities of the enterprise. Data Marts are specific business-related software applications.

data migration. Any process by which data are moved from one physical location, format, and/or platform. In many cases, data migration requires a certain amount of reformatting, modification and in some cases quality checking and control.

data mining. The process of intelligently searching large volumes of data for patterns and relationships. This process is used extensively by marketing firms to extract otherwise unrecognized patterns or relationships in demographic or personal data. Geotechnical data mining attempts to extract relationships between various forms and type of technical data.

data security. A system for any data store to protect the data from unauthorized access, modification or deletion. This system (or set of standards and procedures) also protects the data from hardware failure through a program of regularly scheduled backup operations.

dataset. A collection of related data. A dataset can include digital or physical data, or a combination of the two.

data warehouse. A large-scale, subject-oriented, integrated database which collects the necessary information to support business decisions. Also referred to as a data mart (see which). A data warehouse normally includes a system for the extraction of data, and various reporting functions. Unlike a data mart, which is a highly transaction-based database system, a data warehouse is mainly an archival system that stores information from various databases across the entire data enterprise.

database. A collection or set of related files or relational tables which is created and managed using a database management system (DBMS). A database can contain virtually any type and format of data, and is shared by multiple users in the company, possibly at different geographic locations.

database engine. A software application that stores and retrieves data from a database. Normally, the database engine is included as part of a database management system (DBMS), but it can exist as a stand-alone application.

database management system. The total data management entity, consisting of database engine, data tables, reporting utilities, and related applications.

datatype. A grouping of data based on common characteristics and usage.

DB2. A relational database product, originally developed by IBM for mainframe computers. This DBMS is one of the most powerful database management systems available, and is available on most workstation platforms and operating systems.

DBA. Acronym for DataBase Administrator. This individual is responsible for maintaining the smooth operation of the database system, conducting system backups (and restoring data when necessary), and general maintenance and tuning of the system.

dBASE. Commercial name of one of the first PC-based relational database systems. Originally created by Aston-Tate Corporation, later acquired by Borland, and now an independent company. The table format developed for dBASE® (DBF) is a standard table format for import and export options to and from various PC-based applications.

derived data. Data resulting from a computational process involving data within the database.

distributed database. Databases that are physically stored on more than one computer system, but functions as though it were a single database. If redundant data are stored in multiple locations, updates to one database are automatically updated on the other databases on a regular basis to maintain currency (see which).

DML. Acronym for Data Manipulation Language. This is a generic term for the DBMS programming language that allows a programmer to access and/or modify the contents of the database.

dynamic database. A database which changes periodically due to editing changes, new data additions, and/or data deletions (see also **static database**).

field. The fundamental physical unit of data, made up of one or more bytes of information. Collectively, multiple fields make up a single record in the data table.

FIFO. A data transaction process which posts transactions on a First-In, First-Out order.

filters. A stored set of criteria which specify how a subset of data are extracted from the database. Filters are used during the data retrieval and reporting process to provide the user with only the data they need.

fixed length. Generally refers to the format of a text-formatted export file. In these files, each record is the same overall length, and each data element or field is the same individual length.

flat-file database. Any data file that is not related to, or containing any relational links to other files. Normally, a flat file database is used as a stand-alone collection of data records.

foreign key. One or more columns in a data table which contains a value that matches the primary key in another table.

fourth-generation language. Any type of 4GL (4th Generation (programming) Lanuage). In most cases, these are visual, object-oriented programming languages that allow development of applications at a very high level using natural language commands.

FTP. File Transfer Protocol. A very common way to transfer files from one computer to another, usually over the internet.

function. A small, self-contained software routine that performs a specified operation and returns a resulting value. Functions are normally written in the programming language of the application that they are used in, but can be generic and reusable in nature.

GPS. Acronym for Global Positioning System. An accurate method of determining a location on the Earth's surface using a handheld receiver and multiple satellite transmitters in geosyncronous orbit.

GUI. Acronym for Graphical User Interface. This is a "window" environment with graphical or iconic elements, as opposed to a text-based environment.

HTML. Acronym for HyperText Markup Language. HTML is used extensively in the creation of "pages" of information that can be viewed by any web browser tool.

http. Acronym for HyperText Transfer Protocol. These are a fixed set of commands used during a hypertext link between client and server.

HyperCard. An application development system developed by Apple for the Macintosh computer. One of the first methods of using hyperlinks or hypertext (see which) to create links between different locations in a system.

hypertext. A string of text characters that define a link or pointer to a website, URL, or file location. Most hypertext links are shown in a document with a different font style and/or color, and can be recognized when the cursor is positioned over them changes to a "pointing hand". Selecting a hypertext link will activate the link and move to that page in a browser environment, or to another physical location in a hypertext-linked document.

index. The index is the primary means of linking one data table to another. Indicies in a table can be unique (primary keys) or non-unique (foreign keys).

Informix. A commercial database management system developed by Informix Software. This system runs on most UNIX-based workstations, and is available in client-server versions.

Internet. The internet is a network of connected computers, each of which is identified by a unique IP address (see which).

IP address. All devices on the internet are identified by a unique address, which is used during communications and data transfer to locate a specific computing resource and establish a link for information transfer.

ISAM. Acronym for Indexed Sequential Access Method. This is a common data access method which stores data in a sequential file and keeps track of data locations using an index.

ISO. Acronym for the International Organization of Standards. ISO provides paid subscribers with current approved standards and procedures for virtually every business and industrial field.

JDBC. Acronym for Java Data Base Connectivity, a programming interface that allows applications developed in Java to access a database using SQL commands. This allows platform-independent applications developed in Java to access data from various sources.

Jet. Acronym for Joint Development Technology, a database engine used by both MS-Access, C++, and Visual Basic.

join. The process in a relational database where data elements in two tables are matched on a certain set of conditions, and a third table is created that contains the matching data.

key. A field in record that contains information that identifies that particular record. A primary key is used to uniquely identify that record from all the other records in the table. A foreign key can be duplicated in a table, but each foreign key must have a corresponding entry in another, linked table (where it is the primary key).

keyword. A word in a document or text field which is the target of search operations. Keywords are used extensively in data mining and web searches.

LAN. Acronym for Local Area Network. Normally, a series of compters in a single location (building or group of buildings) connected by physical wiring, routers and switching hubs. Normally a LAN is protected from the "outside" world through the use of physical and software-enabled firewalls.

legacy data. Data from a data management system which is no longer in use, or from which data are being transferred.

legacy system. A data management system which is no longer in use, or which is being replaced by a newer system. In many cases, a legacy system may not have the ability to export data to a format compatible with the new system, requiring the use of text files for data transfer.

mail merge. A common word processing feature that uses information from a database to create customized documents and reports.

merge. The process of joining multiple pieces of information, files, or text to create a concatenated structure.

metadata. Metadata is bascially information about the data, not the actual data itself.

missing data indicator. A value inserted into a data field which indicates that data is missing. In most cases, for numerical data,a value is used that is unlikely or impossible to occur naturally (for example, -999.99)

multidimensional DBMS. A complex data management system that retrieves data in arrays and allows the arrangement of data views in multiple dimensions.

normalization. A process in which a complex database is reduced to the most stable, simple structure possible. This process requires the removal of any redundant atrributes, and unnecessary keys or relationships.

null data. A null data field is empty of all information. This is not to be confused with a blank, which actually has a binary value. Nulls are used to to indicate that a particular data filed is completely empty.

ODBC. Acronym for Open Data Base Connectivity. Used extensively in MS-Dos based platforms to access information from various databases and files.

OLAP. Acronym for On-Line Analytical Processing. Software tools use OLAP to combine summarized data from various tables and sources to provide rapid statistical analysis of complex datasets.

order. A definition that determines how the data in a table are arranged. Changing the order in a table does not necessarily change the physical sort order (see which).

primary key. Any single column or combination of columns in a data table that uniquely identifies each record in a data table. There are no duplicate primary keys in any data table.

RDBMS. Acronym for Relational Data Base Management System. See DBMS

record. A collection of data fields that contain information about a particular subject. Multiple records make up data sets, files, or data tables.

referential integrity. A safety feature of relational databases that ensures that every foreign key in a table has a matching primary key in another table. This feature prevents records from becoming "orphaned". See primary key and foreign key.

replication. A process whereby distributed databases are synchronized through regular duplication of the database on the distributed computers.

report. A general term for output from a data management system. Can range from complexly formatted, formal printed reports to text-only files used to transfer (migrated) data from one database to another.

report writer. Any application which generates a report from a database system. A report writer can be a simple, built-in report generation module, or a stand-alone, sophisticated application with programming and customization options.

RGB. Acronym for Red, Green, Blue color settings.

schema. A complete description of the entire database, either as a graphical representation or in a report format. The schema contains all the information about the tables, structure, data formats, and relational links in the database.

sort. The physical re-ordering of records in a data table into a prescribed order. Similar to ordering or indexing, except that the record order is physically rearranged.

SQL. Acronym for Structured Query Language. Standarized SQL commands are used almost universally to perform various retrievals, replacements, and other important data manipulation operations on a database.

SQL Server. A commercial DBMS product, originally written by Sybase, used for client/server database applications. A proprietary version of SQL Server is also available from Microsoft.

static database. A database which does not change with time, or a project database which is not actively linked to a master database (see also **dynamic database**).

stored procedure. A pre-defined SQL-language program that is stored in the database. The stored procedure can be executed by client computers accessing the database, eliminating the need to have the procedure loaded to every client machine.

table. A table is a collection of individual data elements (columns) or attributes in discreet records, or rows. Collectively, many tables are combined within the DBMS to create a database.

text files. Any electronic file which contains unformatted text-format (ASCII) character data. Text files can include documents, data files, configuration files, etc.

tuple. A row or record in a relational database.

two-phase commit. An editing safety feature used in distributed databases. In the first phase, the edit, or transaction is confirmed as having been received by the database and stored. The second phase does the actual updating of the database.

UDA. Acronym for Universal Data Access, a term used by Microsoft to describe a set of standards for database and file access protocols.

user-defined function. A software procedure or small application that is written or defined by the user in the application language of the database system. UDF's can be shared with other users of the same database.

VSAM. Acronym for Virtual Storage Access Method, a data storage technique developed by IBM for mainframe database applications.

WAN. Acronym for Wide Area Network. Communication between computers on a corporate LAN network in widely separated geographic locations are done using a WAN.

XML. Acronym for Extensible Markup Language. Similar to HTML, XML is an open standard for describing the content of data elements in web pages (while HTML describes how the data elements will be displayed on the web page).

Index

A

abandonment and remediation phase, 41–42
acquisitions and disposals, 42–44
age-dating techniques, 138
angle averaging method, 129
application integration, 60, 160–161, 185
archiving data, 80
ArcView, 58, 151

B

backward integration, 35
basin models, 30
binary data, 99–101, 142
borehole data
    deviation, 123
    environmental corrections, 145
    geometry measurements, 32, 123–131
    geophysical, 34–35, 138–140
    log and, 139–146
    nomenclature, 33
    storage, 140–144
    user involvement in selecting products for, 139
bugs, reporting. *See* problem tracking system

business rules
   for change management, 45
   for data validation, 190
   developing, 37
   for nomenclature, 33
   for source and ownership of data, 29

C

case conversion, 168
CASE tools, 108–109
change management, 45, 81, 88
checkboxes in interface design, 217–218
checkshot survey data, 137
child table, 14
combo boxes in interface design, 217
communication. *See also* user involvement
   checklist, 227–228
   in database manager role, 84–86
   project Web site for, 86
   with users, 47–54, 205–208, 221–222
computational methods for directional survey data, 129–131, 193
computed geotechnical data, 128–129
computing platform selection, 78–80
converting text to numeric data, 167–168
coordinate data, 119–123
core data, 147
customization
   of ArcView, 58
   avoiding, 44–45
   capabilities provided within a DBMS, 67
   of commercial products, 110–112, 222–223
   in exploitation phase, 40
   factors to consider, 73
   for highly interactive applications, 56
   pitfalls, 76, 81
   in reconnaissance phase, 44–45
   in spreadsheet-based databases, 71–72

D

data. *See also* data types
   binary, 99–100, 142
   borehole. *See* borehole data
   checkshot survey, 137
   coordinate, 119–123
   core, 147
   depth-related, 124, 128, 131–132, 193
   derived. *See* derived data
   described, 3
   directional survey. *See* directional survey data
   exploration drilling, 32–33
   geological age, 138
   geotechnical. *See* geotechnical data
   imaging, 27
   line navigation, 136
   log and borehole, 139–146
   logical, 99
   numeric. *See* numeric data
   observed vs. computed, 124–129
   open-hole well logging, 32
   petrophysical, 147–150, 192–193
   relative offset, 129
   remote sensing, 27
   SCAL yield, 147
   seismic trace, 137
   source and ownership of, 29–30, 34, 135, 146
   spatial, 151–152
   stratigraphic, 132–135, 191–192
   surface, 27–29
   well, 28, 33
data assets, inventory of, 38
database administrator (DBA), 65, 83, 85, 183–184
database manager (DBM)
   qualifications, 83–84, 185, 190, 225
   roles and responsibilities, 84–85, 197–198, 200, 206–207, 221
database normalization, 183–188. *See also* data normalization
databases. *See also* data management systems; data normalization; database types; links, database table
   application-driven/independent, 79–80
   comparison, 18–19
   corporate vs. distributed, 59–60, 80

Index

defining function, 48–49
definition, 183
described, 8
dynamic, 55
generic structure, 8f, 9
misconceptions, 4–5
static storage, 55
table links, 14, 15, 77, 104, 172–173

database types
flat-file, 12–13, 18–19, 71
hierarchical, 8f, 9
object-oriented, 72
proprietary, 71
relational, 12–13, 18–19, 70–71
simple, 4–5
spreadsheet-based, 71–72

data dictionaries
availability during validation review, 202
described, 103
importance of, 104–106

data duplication, 90

data export/transfer functions
commercial examples, 170–171
to DBMS format, 173
defining requirements, 56
described, 17
to flat files, 173–175
generalized import procedures and solutions, 176–182
issues, 82
loading from text files, 175–176
primary methods, 74
reformatting, 165–166
for scanned documents, 156

data formats. *See also* data types; numeric data
change management, 45
decimal degrees (DD), 120–122
geotechnical, 90–101
for GIS browser interfaces, 152
for log data, 142
numerical data, 92–94, 96–98
reformatting for internal consistency, 159–168
RP-66 file, 170
rules and recommendations, 92–101
for scanned documents, 155–156
for seismic trace data, 137
standards, 43–44, 56, 58, 82
standards checklist, 226
strategies, 164–165

data import procedures, 176–182
data integration, 29–30, 35, 37
data inventory in reconnaissance phase, 29

data management. *See also* data management systems; data storage; project planning; quality control (QC)
in acquisition and disposal, 42–44
conventions and nomenclature overview, 3
defining project objectives, 48–49, 226
defining user interface, 49, 205–223
early planning, 44, 47
in field-delineation phase, 36–37
flexibility, providing for, 45, 73
for geological age data, 138
log and borehole data, 138–146
misconceptions, 7
nomenclature, 204
petrophysical data, 150
project life cycle, importance of throughout, 21
in reconnaissance phase, 23, 24
subsurface data, 28, 32, 33, 135–136

data management systems. *See also* data management; data models; documentation; interpretive applications; project planning; quality control (QC); relational database management systems (RDBMS)
accuracy issues, 82
automating normalization, 188
change management, 45
commercial examples, 170–171
concepts and terms, 7–12
data transfer issues, 82
DBMS as heart of data management system, 63
described, 8
DIMS, 32
documentation, 202, 226, 228
expandability, 64–65
in the field delineation phase, 36–37
flexibility, providing for, 45, 73
for geophysical data, 34
hardware obsolescence, 81
inter-database data transfer, 172–175
interpretive applications, 74
inventory of data assets, 38
networking capabilities, 65
nontechnical considerations in DBMS selection, 67–70

in preproject planning, 47
project Web site for DBMS status, 86
quality control, building into interface, 214–220
quality control, importance of, 189
rollback option, 107–108
scalability issues, 81
selection criteria, 63–65
standardization in, 80–82
standards checklist, 226
support and maintenance, 40, 44–45, 57–59, 67–70, 75–82, 156, 170, 222–223
tailoring the database to the data, 55
technical considerations, 65–67
types, 70–72
upgrade issues, 64–65, 80–81
user interface design, 49, 205–223
user involvement, 49, 50–54, 139, 208, 221–222, 227–228
visualization tools, 35

data models
accuracy issues, 82
change management, 45, 81
commercial, 75–76
early planning for, 43
extensions, 112
IRIS21, 75
mineral types, accounting for, 148
modifications to, 81, 110
POSC (Petroleum Open Software Corporation), 75
PPDM (Petroleum Public Data Model), 75
proprietary, 45, 77–78, 88
scalability, 88
selecting, 75–76
and spatial metadata, 152
standards, importance of, 57, 80

data normalization. *See also* database normalization
automating processes, 187–188
concepts, 184
described, 10, 183
of log data, 145–146, 195
methods, 185–188
procedures, 37

data quality control. *See* quality control (QC)

data redundancy, 90

data reformatting, 160–168, 178–182

data retrieval, 16, 56

data storage. *See also* data formats; derived data
archival, 80
deleted records file, 107–108
digital documents, 153–157
directional survey data, 130
in exploration-development phase, 34–35
in field-delineation phase, 37
formation testing data, 146
history files, 107–108
log and borehole data, 140–144
original vs. derived data, 90–91, 105–106, 130
requirements definition, 55, 59–60
scalability issues, 81
SCAL yield data, 147
scanned documents, 155–157
standards, 71
survey information, 136–137
using DLIS format, 142
volume impacts, 32, 34–35, 37, 40, 78, 148
zones and layers data, 133

data types. *See also* data formats; geotechnical data
Boolean, 99
changes to, 178
consistency, 57
definitive and project, 82
in exploration-development phase, 31–35
in field delineation phase, 36
geophysical, 137
geotechnical, 91–101
in proprietary data models, 78
validation of, 190–191

data utilization, defining requirements, 56

data validation. *See also* numeric data; quality control (QC); standards
auditing, 198
building capabilities into interface, 214–220
of data dictionary, 104
data dictionary, availability of, 202
date controls, 219
definition, 189
directional survey data, 193
duplicate or redundant data, 90
editing methods, 197–198

form-based, 214–220
general methods, 190–193
geostatistical methods, 193–196
import and export considerations, 220
quality control, 89, 201–203
reporting problems, 198–200
reviewing validation methods, 203
rules and recommendations, 89–91
rules database, 202
stratigraphic, 191–192
testing of validation rules, 196
tools, 196–197
training of staff to prevent errors, 90
date and time data, 97–99, 135–138
date controls in interface design, 219
DBA (database administrator), 65, 83, 85, 183–184
DBM (database manager). *See* database manager (DBM)
DBMS, see *under* data management systems
deleted records file, 107–108
depth points, duplicate, 193
depth-related data, 124, 131–132, 193
derived data
  described, 105
  petrophysical, 149–150
  storing, 90–91, 105–106, 130
designing the user interface. *See* user interface design
development phase, 38
deviation data format variations, 125–127
DHI (direct hydrocarbon indicators), 27–29
digital document storage, 153–157
Digital Log Information Standard (DLIS), 142
DIMS data management system, 32
direct hydrocarbon indicators (DHI), 27–29
directional survey data
  computational methods, 123–130, 193
  measured data, 124–125
  methods of gathering, 123
  miscellaneous information to store, 127–128
  observed vs. computed data, 124–129
  storing computed results, 130

validation of, 193
disposals and acquisitions, 42–44
distributed databases, 61
documentation
  database design, 228
  data dictionaries, 104–106
  data import formats, 161
  data validation rules and procedures, 202–203
  of modifications, 81
  planning for, 58
  preplanning, 228
  project objectives, 228
  standards checklist, 226
  user instructions and programming support, 58, 69, 228
document management systems, 30, 33, 38, 42, 155–157
documents, scanning, 154–157
dogleg severity, 129
drilling operations, 32
drop-down boxes in interface design, 215
duplication of data, 90

E

exit strategy, 38
expandability of a DBMS, 64–65
exploitation phase, 39–40
exploration-development phase, 31–35
exploration drilling data, 32–33

F

field-delineation phase, 36–37
fields, data, 10–11
fields, key and index, 11, 13, 77, 104
flat-file databases, 12–13, 18–19, 71
flat-file data transfer, 173–175

fluid saturations, 148
foreign keys, 13, 77
form-based validation, 214–219
forward modeling, 35
functions, data export/transfer. *See* data export/transfer functions

G

geological age data, 138
geophysical data. *See under* geotechnical data
GeoQuest products (Schlumberger), 170–171
geotechnical data. *See also* data management; data management systems; GIS (geospatial information system) browser; numeric data
   acquisition and interpretation, 34–35
   change management, 45
   computed, 128–129
   coordinate data, 120–123
   database application selection for, 72–80
   data management of oil and gas assets, 21–45
   data model selection/development, 75–78
   depth-related data, 131–132
   described, 119
   in development phase, 38
   digital document storage, 152–157
   directional survey data, 123–131
   in disposals, 43
   in exploration-development phase, 21
   formats, 91–101
   geochemical surveys, 28
   geophysical data, 136–137
   GIS interfaces, 151–152, 212–213
   log and borehole data, 34–35, 124–125, 138–147
   major categories, relative importance of during project life cycle, 23f, 31f, 36f
   petrophysical data, 147–151
   reformatting, 161–164, 167–168
   spatial data and GIS systems, 151–152
   standards, fitting to designed, 44
   stratigraphic data, 132–135
   time-related data, 135–138
   types, 91–101
   user involvement, 50, 54, 70
GIS (Geographic Information Systems) systems, 151–152
GIS (geospatial information system) browser interface
   in asset review and data delivery, 44
   data formatting issues, 152
   in facilities planning, 37
   for information integration and visualization, 29–30
   in interface design, 208, 212–213
   usefulness to geoscientists, 151–152
graphical user interface (GUI), 49, 211–212

H

hardware obsolescence, 81
header information in data reformatting, 162–163
help desk software, 200, 228
hierarchical databases, 71
hierarchical indexing system for scanned documents, 155
histograms, 194, 195
history files, 107–108

I

images as binary data type, 100
indexing
   metadata, 25–26, 29
   scanned documents, 155
   versus sorting, 166–167
industry-standard data formats, 43
integration, data, 29–31, 35, 37
integration of applications, 60, 160–161, 185
interface design. *See also* user interface design

checkboxes, 217–218
combo boxes, 217
date controls, 219
drop-down boxes, 215
GIS browser interface, 208, 212–213
numeric data validation, 218
radio buttons, 217
spinboxes, 218–219
surveys for, 205–206
user involvement, 205–208, 221–222

interpretation time, 3

interpretive applications
in backward integration, 35
data types and, 137, 142
in exploitation phase, 40
format standardization, 43–44, 45
GIS interface, 30
hardware considerations, 59
linking to, 74
operating systems and, 62
single-vendor solution, 66f
vendor products, 79, 170–171

IRIS21 data model, 75

K

key fields
described, 11, 13
in proprietary data models, 77
relationships defined in data dictionary, 104

keywords, defining, 155, 157

Kick-Off Point (KOP), 124f

L

Landmark OpenWorks (Halliburton), 171

LAS-format log data files, 140

life cycle stages, overview of, 22, 23f

line naming convention, 134–135

line navigation data, 136

links, database table
defining, 77
documented in data dictionary, 104
many-to-one relationship, 15
ODBC and SQLNet links to tables, 172–173
one-to-many relationship, 15
one-to-one relationship, 14

Linux, 62

list boxes in interface design, 216

log analysis, 150

log and borehole data, 138–146. *See also* borehole data; log data

Log ASCII Standard (LAS) file, 18, 71, 140, 140–142, 143f

log data
and borehole data, 138–146
data management problems, 145–146
LAS-format log data files, 71
normalization procedures, 195
petrophysical data and, 148–150
in reconnaissance phase, 32

log header components, 139–140

logical data, 99

Log Information Standard (LIS), 142

M

maintenance. *See* support and maintenance

manuals. *See* user instructions

many-to-one relationship, 15

map-based (GIS) interfaces, 212–213

map projection constants, 44

mean sea level (MSL), 124f

measured depth, 124

meets and bounds, 123

menu-based interface design, 208–211

metadata
collecting, 30
geographic, 25
indexing, 25–26, 29
for scanned documents, 155, 157
spatial, 152, 212

mineralogical volumes, estimation of, 148
minimum radius of curvature method, 129
modeling, forward, 35
modifications. See change management; customization

N

navigation data, 136
net pay values, 149–150
nomenclature
    advisory groups, 227
    in data management, 3–4
    in exploration-development phase, 33
    stratigraphic, 53, 134–135
normalization. See data normalization
numeric data
    converting from text, 167–168
    converting to special text formats, 168
    in data dictionary, 104–105
    latitude and longitude, 120–122
    rules and recommendations, 92–98
    validation, 191
    in validation interface design, 218

O

objectives, defining, 48–49, 226
object-oriented databases, 72
object-oriented programming (OOP), 211–212
ODBC and SQLNet links to tables, 172–173
one-to-many relationship, 15
one-to-one relationship, 14
open-hole well logging data, 32
OpenSpirit consortium, 171
operating systems
    and interpretive applications, 62
    obsolescence of, 81
    personal computer (PC), 61–62
    requirements, 61–62
    standards, importance of, 57
Oracle, 64
ownership of data
    establishing rules for, 29–30
    geophysical data, documenting, 34
    log data, 146
    stratigraphic tops data, 135

P

parent table, 14
PC (personal computer) operating systems, 61–62
permeability, determination of, 149
Petroleum Open Software Corporation (POSC), 72, 75
Petroleum Public Data Model (PPDM), 75
petrophysical data, 147–150, 192–193
pick lists in interface design, 216
pilot data files, 178
planning. See project planning
platform selection. See computing platform selection
policies, standards, procedures, and guidelines (PSPG), 201
porosity, 148
portability of a DBMS, 88
primary key, described, 13
probability distributions, 194
problem tracking system, 54, 198–200, 201, 228
project planning. See also data management
    computing platform selection, 78–80
    coordination as prime role of database manager (DBM), 85–86
    database application selection, 72, 76
    database management system selection, 64–70
    database tailored to data, 55–57
    data model selection, 75
    defining objectives, 48–49, 226
    digitizing documents, 153–157
    flexibility, providing for, 45, 73

Index   261

hardware considerations, 59–62, 81
interpretive applications, linking, 74
problem tracking, 54, 198–200, 201, 228
project Web site for communication, 86
scalability considerations, 81
support and maintenance considerations, 40, 44–45, 57–58, 67–70, 75–82
surveys, 51–54, 206
user interface design, 205–223
user involvement, 47–54, 205–208, 221–222, 227–228
proprietary databases, 71, 76
proprietary data models, 77–78
proprietary solutions. *See* customization

Q

quadrant and bearings, 125
quality assurance. *See under* quality control
quality control (QC). *See also* standards, importance of
   building into the interface, 214–220
   at database point of entry, 89
   in data loading and input, 169
   in data management project, importance of, 189
   in data validation, 89, 201–203
   and definitive data, 82
   documentation, 201
   limits in data tables, 184
   methods, 201
   quality assurance, 60, 82, 201, 203
queries, format considerations for, 94–95
query-by-example (QBE) interface, 16

R

radio buttons in interface design, 217
radius of curvature method, 129
reconnaissance phase, 23–30
records, database, 10
redundancy of data, 90

reformatting data, 161–168, 178–182
relational database management systems (RDBMS)
   advantages, 70–71, 104
   and CASE tools, 108
   concepts, 12–19
relationships, database table. *See* links, database table
relative offset data, 129
remote access of databases, 59f
remote sensing data, 27
report formatting, removal of, 163
reporting functions, 18
reserves/resource estimates, 30–31, 38
reservoir rock computation criteria, 149
retrieval of data
   defining requirements, 56
   use of Structure Query Language (SQL), 16
rollback option in a DBMS, 107–108
RP-66 data file, 170

S

scalability of DBMS, 81, 88
SCAL yield data, 147
scanning contractor, evaluating, 154–155
scanning documents, issues with, 154–157
schema, described, 9
security, database, responsibility of database administrator (DBA), 85
SEG-Y data storage standards, 71
seismic trace data, 137
shapefiles, 152
shotpoint and line information, 136
software obsolescence, 81
sorting vs. indexing data, 166–167
source of data, 29–30, 34
spatial data, 151–152
spatial metadata, 152, 212
special core analyses (SCAL), 147

spelling in database text strings, 94
spinboxes in interface design, 218–219
splicing logs, 145
spreadsheet-based databases, 71
SQLNet links to tables, 172–173
SQL (Structure Query Language), 16, 185–186
standards, importance of. *See also* support and maintenance
    in acquisitions and disposals, 43–44
    and change management, 81
    for data formats, 43–44, 56, 58, 82
    in data loading and interpretation process, 43
    in data models, 57, 80
    in data storage, 71
    within DBMS programming capabilities, 58, 67
    development of as role of DBM, 84
    in documentation, 58, 226
    in indexing digitized documents, 155
    key points summary, 225–227
    in project life cycle, 37, 44–45
    throughout project planning, 57
    in quality assurance, 60, 82
    reformatting data, purpose of, 160
    upgrade and migration issues, 80–82
    in well naming, 33
statics and velocity data, 137
storage of data. *See* data storage
stratigraphic data, 132–135, 191–192
stratigraphic nomenclature, 53, 134–135
Structure Query Language. *See* SQL (Structure Query Language)
subsurface samples, 28, 32
support and maintenance
    across multiple time zones, 67–70
    for Apple Macintosh systems, 62
    of commercial products by vendor, 170
    considerations
        in computing platform selection, 78–80
        in data model selection/development, 75–78
        in DBMS selection, 65, 67–70
        in project planning, 57–58

    of customized products, 222–223
    custom solutions, 40, 44–45
    proprietary data models, 77
    scanned documents, 156
    staffing considerations, 40
    standards, importance of, 80–82
surface data, 27–29
survey data, computational methods for, 129–131
surveys
    geophysical survey data, storage of, 136–137
    for project planning, 51–54, 206
    to solicit user feedback, 227
    surface geochemical, 28
    for user interface design, 205–206

T

table-based validation, 220
tables, database. *See also* links, database table
    described, 9
    generic, 112–117
    in relational databases, 13–18
tangential method, 129
tar deposits, 28
technical documentation. *See* documentation
technical support. *See* support and maintenance
test data, geophysical, 28, 33, 37
testing
    data import or conversion routines, 175, 178
    DBMS modifications, 85
    integrity of backup procedures, 85
    log data systems, 139
    user interface prototypes, 207
    validation rules, 196
    of vendor demonstration products, 213
    when automating normalization, 188
text conversion, 167–168
time and date data, 97–98, 135–138

tools for database applications, 109–110. *See also* CASE tools
tops and bases, 134–135
training, 70, 90
true vertical depth subsea (TVDSS), 124f
true vertical depth (TVD), 124f, 128

U

ULTRIX, 62
Universal Transverse Mercator (UTM) projection method, 122
UNIX, 62
upgrade options for a DBMS, 64–65, 80–81
user instructions, 58, 69, 228
user interface design
  commercial interfaces, customizing, 205–223
  GIS (map-based) interfaces, 212–213
  GUI (graphical user interface) design, 211–212
  individual interviews, 206–207
  menu-based interface design, 208–211
  object-oriented design, 211–212
  programmatic solutions, 221
  prototypes, 207
  standards checklist, 226–227
  survey strategy, 206
  user involvement, 49, 205–208, 221–222
  validation considerations, 214–220
user involvement
  checklist, 227–228
  critical for a successful project, 49
  in project life cycle, 50–54
  in project planning, 47–54, 227
  in selecting log and borehole data products, 139
  in user interface design, 205–208, 221–222

V

validation of data. *See* data validation
validation rules database, 202
velocity and statics data, 137
vertical offset, 129
vertical seismic profile (VSP), 137
views, 17
volume impacts on data storage. *See under* data storage

W

well-centric subsurface data, 33
well data, 10f, 11f, 21, 28
wide area network (WAN), 59–60
workstation operating systems, 62

Other titles offered by PennWell...

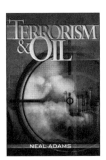

Terrorism & Oil
by Neal Adams
193 pages, softcover
$34.95 US/CAN
$49.95 Intl
ISBN: 0-87814-863-9

Drilling Technology in Nontechnical Language
by Steve Devereux
337 pages, hardcover
$64.95 US/CAN
$79.95 Intl
ISBN: 0-87814-762-4

Well Logging in Nontechnical Language, 2nd Edition
by David E. Johnson and Kathryne E. Pile
289 pages, hardcover
$64.95 US/CAN
$79.95 Intl
ISBN: 0-87814-825-6

Petroleum Production in Nontechnical Language
by Forest Gray
288 pages, hardcover
$64.95 US/CAN
$79.95 Intl
ISBN: 0-87814-450-1

Computer-Assisted Reservoir Management
by Dr. Abdus Satter, Jim Baldwin and Rich Jespersen
278 pages, hardcover
$84.95 US/CAN
$99.95 Intl
ISBN: 0-87814-777-2

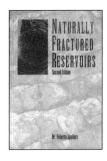

Naturally Fractured Reservoirs, 2nd Edition
by Dr. Roberto Aguilera
521 pages, hardcover
$99.95 US/CAN
$114.95 Intl
ISBN: 0-87814-122-7

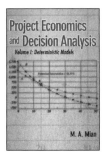

Project Economics & Decision Analysis, Vol. 1 Deterministic Models
by M.A. Mian
397+ pages, hardcover
$74.95 US/CAN
$89.95 Intl
ISBN: 0-87814-819-1

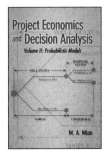

Project Economics & Decision Analysis, Vol. 2 Probabilistic Models
by M.A. Mian
490 pages, hardcover
$84.95 US/CAN
$99.95 Intl
ISBN: 0-87814-819-1

To purchase a PennWell book...

- Visit our online store www.pennwell-store.com, or
- Call 1.800.752.9764 (US) or +1.918.831.9421 (Intl), or
- Fax 1.877.218.1348 (US) or +1.918.831.9555 (Intl)